WORKING WITH CHEMISTRY: A LABORATORY INQUIRY PROGRAM

DONALD J. WINK AND SHARON FETZER GISLASON
University of Illinois at Chicago

JULIE ELLEFSON KUEHN
William Rainey Harper College

W. H. Freeman and Company
New York

Acquisitions Editor: Jessica Fiorillo

Text Design and Composition: York Production Services

Cover Design: Blake Logan

Production Coordinator: Susan Wein

Manufacturing: RR Donnelley & Sons Company

Cover Images; counterclockwise: Mark Joseph/Tony Stone
Phil Schermeister/Corbis
Roger Ball/Stock Market
Josh Mitchell/Tony Stone (Lab coats in background)

Cataloging-in-Publication Data available from the Library of Congress.

Printed in the United States of America

ISBN: 0-7167-3549-0

First printing 1999

W. H. Freeman and Company
41 Madison Avenue, New York, NY 10010
Houndmills, Basingstoke   RG21 6XS, England

Working with Chemistry:
A Laboratory Inquiry Program

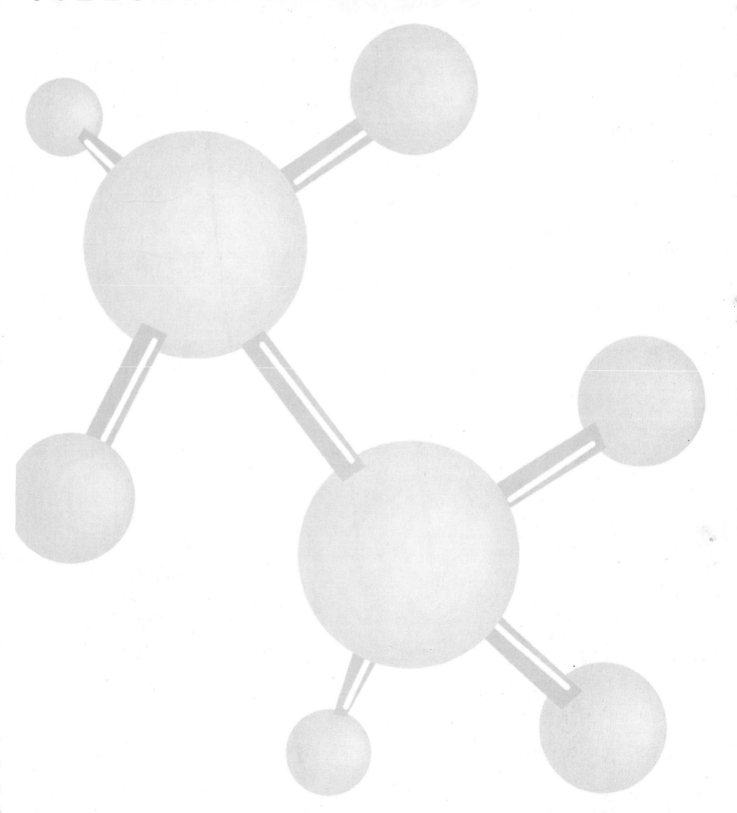

CONTENTS

EXPERIMENT GROUPS

WORKING WITH CHEMISTRY LABORATORY SEPARATES

The experiment groups listed below are available as Working With Chemistry Laboratory Separates. The Separates are self-bound, self-contained exercises. They are 8½ inches by 11 inches in size and are punched for a three-ring notebook. They can be ordered in any assortment or quantity. Order through your bookstore, specifying the ISBN number and title.

07167-9456-X	Methods of Inquiry and Measurement: Float an Egg
07167-9408-X	Metal Ions and the Blood: Treat Iron Deficiency . . . and Overload
07167-9407-1	Analysis of Complex Solutions: Bloodwork Quick with a "Finger Stick"
07167-9406-3	Acid–Base Titration and the Global Carbon Cycle: Predict Effects of Rising Levels of CO_2
07167-9405-5	Buffers and Life: Save a Cardiac Arrest Patient
07167-9404-7	Measurements of Chemical Reaction Rates: Clean Up Waste Water
07167-9403-9	Electrochemical Glucose Monitoring: Construct and Use Your Own Electrode
07167-9401-2	Heat Capacity: Design a Fireproof Safe
07167-9402-0	Chemical Engineering Process: Manipulatng the Outcome of Reactions
07167-9400-4	Ecological Element Cycles: Judge Levels of Environmental Disturbance

PREFACE

THE IDEA OF *WORKING WITH CHEMISTRY*

We are pleased to introduce you to our general chemistry laboratory program, *Working with Chemistry (WWC)*. This program started from the idea that we could build a program for all students that relates general chemistry to people who use chemistry on a regular basis, even if they don't have "chemist" as part of their title.

We do this while covering essential principles taught in general chemistry. Here are just three examples of how standard general chemistry questions are recast in our program:

1. "How does reaction stoichiometry matter in this lab?" becomes "What stoichiometry is used when a field ecologist needs to determine if a soil is rich in nitrogen?"
2. "What are the characteristics of an acid–base buffer?" becomes "How can a buffer be restored to a patient after cardiac arrest?"
3. "What is the heat capacity of a metal?" becomes "Which of these substances is best suited to protect valuable documents from fire?"

The ideas that underlie these labs are drawn from the education research and policy literature of the last ten to twenty years. The way to forge connections between professional and basic science training has already been enunciated in an NSF report*:

- Form instructional teams of specialists across traditional departmental lines.
- Teach basic science and mathematics through the vehicles of real-world technical problems and industrial scenarios.
- Encourage the collaboration of industrial and academic scientists and technicians.

The WWC materials incorporate these connections in several ways. First, we build upon a discussion of the importance of chemistry in a professional environment with a researcher in that area. Second, each Experiment Group provides instruction in one or more basic ideas or techniques applied to a genuine scenario. Third, collaborative work built around individual responsibilities is a component of each experiment.

GOALS AND STRUCTURE OF *WORKING WITH CHEMISTRY*

The *Working with Chemistry* program has a primary goal that is present in any general science course: understanding how chemistry gathers evidence and solves problems. To accomplish this, we have two additional goals that will serve this primary goal. First, students should develop a mastery of techniques and ideas so that they can gather and interpret data independently. Second, the students should be able to articulate, with ref-

---

*National Science Foundation. *Gaining the Competitive Edge: Critical Issues in Science and Engineering Technician Education* (NSF 94-32). National Science Foundation: Washington, DC, 1994.

erence to the written materials and personal experience, how someone in a real work environment might find a particular technique or idea useful in solving a problem.

Note that these goals are *not* vocational. A general chemistry course should not be viewed as a place to learn particular technical skills. But *Working with Chemistry* does aim to give students strong academic skills that will be remembered when it is time to learn actual workplace skills. This is why we feel the WWC program is also important to students in normal academic tracks. For them, an enhanced understanding and enjoyment of chemistry should aid in the development of a clear understanding of the importance of basic science.

Each *Working with Chemistry* experiment group is written to cover multiple weeks of experimentation. Students read the scenario before the first week's work, and this anchors the instruction for the entire experiment group. The scenario returns as the focus of the important problem they are asked to address in the final experiment.

> **Skill Building Labs** show students how to use a technique. These labs are largely drawn from existing methods commonly taught in the General Chemistry Laboratory.
>
> **Foundation Labs** introduce the use of the technique to solve a chemical problem. The focus is on well-defined material, and students are challenged to produce some of their own procedures to solve the problem.
>
> **Application Labs** return to the professional scenario we have identified. They are more open in their inquiry style.

Within this structure there is some flexibility for the instructor. It is possible to omit either the skill building or the application lab if only two weeks are available for a given topic.

THE REVELANCE OF *WORKING WITH CHEMISTRY*

The sheer number of majors that require general chemistry (25 at our schools alone) make it one of the most notorious courses on any campus. Students struggle with the inherent difficulty of the cumulative learning required in chemistry and their need for chemistry to prepare themselves for a distant employment goal. Orville Chapman of UCLA has clearly (and bluntly) summed up how this struggle compromises the effectiveness of a traditional fact- and computation-focused curriculum:

> We have no hope of expanding our clientele with our present structure. The structure is not sound. The message from 80% of college students comes in loud and clear. Chemistry without people, economics, and policy is irrelevant. We choose to be irrelevant; they ignore us.*

This call for relevance can be and is addressed in curricula that make students more aware of the importance of chemistry to societal problems, especially those of the environment. Another approach is to bring more contemporary research problems into the course. Both have the advantage of making "newspaper" and "community" chemistry accessible to the student. However, for students more focused on a professional degree there needs to be something more: a clear demonstration that, yes, chemistry does matter to *employment*.

In our case, we felt that the laboratory is the most attractive place because we can both develop and show applications for general principles and techniques. The *Working with Chemistry* labs allow students to experience how the solution of real problems by people in all walks of life requires a thorough understanding of general chemistry principles.

---

*National Science Foundation. *Innovation and Change in the Chemistry Curriculum* (NSF 94-19). National Science Foundation: Washington, DC, 1993.

INTERDISCIPLINARY INSTRUCTION IN *WORKING WITH CHEMISTRY*

The WWC program includes a strong interdisciplinary basis that, we hope, provides meaningful issues, examples, and experiences from outside of chemistry. Certainly as authors we have been enriched in the course of discussions in which we learned, for example, that current, not voltage, is a primary means for assaying concentration or how the simple question of fluid flow, combined with simple kinetics, became a key to a successful synthetic procedure. The same enrichment occurs when students construct a calibration curve to assay albumin and compare results about their "patients."

As suggested earlier, we worked hard to structure the interdisciplinary approach of WWC so that it did not compromise the core subject, chemistry. That is critical in classes where students often do not have the same skills before the class starts. We carefully wrote the skill building and foundation labs to bring all students to a common understanding of how the chemistry of the system works before we ask them to think about using chemistry in another area.

INQUIRY, TEACHING, AND LEARNING IN *WORKING WITH CHEMISTRY*

Each experiment group connects to one or more of the *concepts and skills in general chemistry*. This keeps the "standard" lab goal of supporting basic fact and concept learning in place. Also, it provides adequate training during the first two weeks of an experiment group to allow students to appreciate fully the relevance of the applications in the third week. When we address the problem of providing a *working environment*, we do not just want to tell students how chemistry is used by professionals. We want to make them feel the actual environments.

In our implementation of a *guided-inquiry* format, students are responsible for collecting and interpreting data in order to answer some testable question. We do not expect students, for example, to be able to design an acid–base titration experiment on the first day they use a buret. But within a short time span, they will be expected to carry out the procedure to meet certain parameters. In some cases, students determine how to collect the data. They also decide if their data are acceptable and, if not, they have the opportunity to make adjustments and try again. Finally, they are expected to make connections between the scenario, the data they collected, and the chemical principle underlying the experiments.

Professionals are expected to be able to do *collaborative work* from a base of independent competence. Students have ample opportunities to develop individual laboratory skills and to analyze and communicate their own results. However, each experiment group also includes a team component. This may mean collaborating on a procedure, sharing group data, or preparing valid samples for one another.

ACKNOWLEDGMENTS

There are only three listed authors of this lab manual, but in fact it represents the collaboration of more than a dozen people who ably and enthusiastically contributed to the program. But several people have played a role in shaping such large sections of the work that they deserve general thanks.

Perhaps the most important co-worker for us is Mr. William Haney, Coordinator of the UIC Undergraduate Chemistry Laboratories. Only Bill could ensure us that the kinds of experiments we planned would be practical in real teaching environments. More important, he had a special eye for understanding that what we wrote might not be what the students would understand, and his feedback from classroom observations and conversations with teaching assistants has been a big part of the WWC instructor's manual.

With the exception of the introductory lab group, each WWC experiment group has a scenario written in conjunction with a professional faculty member. Their work was not restricted to just "their" experiment, however. Group discussions and planning by all WWC authors and professional faculty were places where everyone bounced ideas freely and thus they all deserve some general thanks for their work. These faculty are John Lussenhop (Biological Sciences—Groups D and J), John Regalbuto (Chemical Engineering—Groups F and I), John Fitzloff (Pharmacy—Group B), Chuck Woodbury (Pharmacy—Group C), Janean Holden (Medical–Surgical Nursing—Group E), JoanAnn Radtke (Medical Lab Sciences—Group G), and David France (Mechanical Engineering—Group H).

Drs. Luke Hanley and Audrey Hammerich of the UIC chemistry department also played an internal role as editors of some of the experiment groups. Colleagues who have also taught these same students enabled us to check that our feel for students' prior and developing knowledge was reasonable. Similar outside perspectives were provided by Dr. Allan Smith of Drexel University and Dr. Baird Lloyd of Miami University Middletown.

The final group of people who contributed in detail to the structure of the WWC labs includes several dedicated colleagues at other schools. First among these are six faculty at Chicago-area community colleges, who helped in planning, in evaluating, and in serving in the initial implementation of these labs. They are Drs. Helene Gabelnick and Dennis Lehman (Harold Washington College of the City Colleges of Chicago), Susan Shih and Dan Fuller (College of Du Page), Jerry Maas (Oakton Community College), and Barbara Weil (William Rainey Harper College). In addition, several people have tested these labs over the last year on their own campuses through our initial download option. Their comments on what happens when the WWC labs went to distant sites have helped in particular and in general.

And, of course, the students at our campuses were the ones who first experienced these labs. Their experiences, in part observed and reported upon by Prof. Elizabeth Whitt (Iowa) and Ms. Han Mi Yoon (UIC), were the primary piece of evaluation data for the shaping and reshaping of the WWC labs. A select group also served as special testers, and we proudly share their data in the student results section of the instructor's notes.

Special thanks are due to the many instructors who reviewed this material before publication. Robert Allendoerfer, State University of New York, Buffalo; Susan Bates, Ohio Northern University; Stacy Lowery Bretz, University of Michigan—Dearborn; Dave Bugay, Kilgore College; Dave Cohen, J. Sargeant Reynolds Community College; Alan Cooper, Worcester State University; David Frank, California State University, Fresno; Robert Hammond, University of Maryland, College Park; Rick Hartung, University of Nebraska, Kearney; Sara Iaccobucci, Tufts University; Sanda Lamb, University of California, Santa Barbara; Eric Mechalke, Casper College; Lynne O'Connell, Boston College; Norb Pienta, University of North Carolina, Chapel Hill; Laurence Rosenheim, Indiana State University; Theodore Sakano, Rockland Community College; Marcy Towns, Ball State University; MaryJon Whittemore, Forney High School, Texas; Marie Wolff, Joliet Junior College; Corbin Zea, Creighton University; Gail Zichittella, Cheektowaga Central High School, New York.

We want to thank the many people at W. H. Freeman who helped make this first edition possible, with special thanks to Chris Rugger, Todd Elder, Kimberly Manzi, Jessica Fiorillo, and Michelle Julet. Finally, we acknowledge the support of the National Science Foundation in getting this project underway and the key connection made by Brian Coppola of the University of Michigan at his Day 2-to-40 symposium in May, 1997.

INTRODUCTION

I. PROGRAM COMPONENTS

A. Introduction to the *Working with Chemistry* Program

During this semester you will use one or more experiment groups developed as part of a program called *Working with Chemistry* (WWC). This program is designed to show how many general chemistry principles are used by people in many different fields. This is a large group of people, ranging from field ecologists to chemical engineers, and including many health professionals as well.

The design of a WWC experiment group is simple. During the experiment group you will learn techniques that you will then apply in the study of a particular problem. We have drawn the "problems" from the work that is done by chemical professionals, and these people have helped us design each experiment group from the ground up. During an experiment group, you will have more and more responsibility for designing the actual procedures. Thus, the first week is the "Skill Building Lab," the second week is the "Foundation Lab," and the third week is the "Application Lab."

Working with Chemistry is a program developed by chemists working with faculty from other departments who use chemistry in their everyday work. They may be biologists, pharmacists, or engineers. The purpose is not to train you how to be one of these persons, but to show you how professionals use the principles of general chemistry in their work. We hope you will develop a better understanding of how chemistry is used today.

Of course, professional chemists usually do not work in an environment where the same procedures can be used every time. They usually have to find out something about a patient, for example, and then design a response. This leads us to the second important feature of the WWC labs: They are based on your own responsibility to inquire, interpret, and act upon results.

Finally, chemical professionals may be individually responsible for their own work, but they do often work as members of teams. When directed, you will therefore assemble a set of data with a group of students. This pooled data will allow you to carry out your individual work in a faster and more accurate manner.

B. Forming and Working in Groups

If you have never worked in a group before, it can feel threatening the first time because you may not know what is expected of *you* as an individual. The process is actually very simple, and as you become familiar with group work within a chemistry lab environment, you will realize that it is the same process you often use to approach your everyday tasks. *Each person in the group is important, because each person has a definite part to contribute to the completion of the experiment.* First you must understand what the group is expected to complete by the end of the lab period. This information is contained in the experimental description and procedure, which should be read before coming to lab. In reading through the experiment, you should understand the reasons (or theory) behind the procedure as well as how to perform the necessary measurements. It helps to highlight important information in each experiment before you come to lab.

After your group has listed all the tasks that must be accomplished, these should be distributed among the individual group members. For the experiment to be successfully completed, each group member should clearly understand the assigned task and how it is to be completed. It is always a good idea to *discuss this as a group first* before going ahead with individual tasks. The group discussion should include the degree of accuracy required for all measurements, a review of laboratory techniques not used in a while, an investigation of new laboratory procedures, and any cautions to be observed within the performance of the experiments. Your group must also organize data collection.

As an example of how the group might work, let's see how these general directives apply to a procedure to study the relative amounts of materials needed in a chemical reaction. If a group of 4 must report results for two solutions, then two students should study each solution. All members would work with the same volume of the solution however. That way everyone has the same volume to compare.

Your group may prefer a different division of labor that will accomplish the experimental goals just as well as this one: The essential point is that all work is done and reviewed in the time allotted, with all the members of the group participating. Important things to notice and record are the names of your group members and the letter (or code) of any unknowns assigned to each person. Take time to record this information at the beginning of the lab period.

Members of a well-functioning group should be able to ask questions within their group, provide direction to members in need of such, and both give and accept critical analysis of the quality of their measurements. At the end of the experiment you may be asked to evaluate the group's performance. This should improve both individual and group understanding and performance.

II. Safety in the Chemical Laboratory

A. Chemical Hazards: Identification and Management

Each person in a laboratory is responsible for maintaining a safe environment. Therefore, you must be aware of any hazards associated with the chemicals used in the experiments. You must always wear appropriate safety equipment, including safety goggles, while working in the lab. It is also your responsibility to follow the safety rules and regulations established by your institution and your laboratory instructor.

Some of the chemicals used in some of the WWC experiments have been identified as harmful or potentially harmful to humans. The chemicals may be corrosive, chemicals that destroy living tissue and equipment on contact; flammable or combustible, substances that give off vapors that can readily ignite under usual working conditions; irritants, substances that have an irritant effect on skin, eyes, respiratory tract, and so on; toxic, substances that are hazardous to health when breathed, swallowed, or in contact with skin; carcinogenic, substances that are known or suspected to cause cancer; mutagenic, chemical or physical agents that cause genetic mutations; and teratogenic, substances that cause the production of physical defects in a developing fetus or embryo. To minimize your risk, it is essential you handle these materials properly, carefully, and while wearing appropriate protective equipment. In fact, a good general rule for all chemicals, even those considered nonhazardous, is to avoid contact with skin, clothing, and eyes. You must also avoid breathing vapors.

To assist you in understanding the hazards associated with specific chemicals, each WWC experiment contains Cautions to alert you to potential hazards. Additional information on each chemical is available on its Material Safety Data Sheet (MSDS). Every MSDS contains specific information including physical and chemical characteristics, health effects, fire and explosion data, reactivity hazard data information, health hazard data, and precautions for spills and cleanup. MSDS sheets should be available from companies that supply reagents.

B. Disposal of Chemical Substances

All chemical waste must be disposed of properly, generally in appropriately labeled containers. Your institution has a chemical hygiene plan that includes procedures for waste disposal. Follow all disposal procedures as outlined by your instructor. Be sure to also clean all glassware, equipment, and your work area before leaving the laboratory.

III. INSTRUMENTS FOR CHEMICAL MEASUREMENT

A. Handling Chemicals

As outlined in the previous section, all chemicals must be handled carefully with attention to personal safety. In addition, the use of laboratory chemicals, whether pure solids or solutions, requires care to be certain that little is wasted and that the purity of the substances and solutions is protected.

Before you use any chemical, review the procedure to determine how much you will need. You should *never* draw material directly from a common vessel. Use a clean, dry vessel to obtain the material for use by yourself or your lab group. Then, all measurements of the amount of the material can be made from this portion. That way, if your technique contaminates the sample, only a small amount, not a whole bottle, is affected.

If you find that you have excess material left over after you have prepared your experiment, then treat this material as waste. You do not know if it has become contaminated in handling, so it should never be returned to the common stock.

B. Measurement by Mass

Measurement of the mass of a substance is a critical part of many experiments in general chemistry. It is essential that this be done in a precise and accurate manner. The *precision* is dictated by the instrument. We typically refer to a balance by the last digit that it provides on a gram (g) basis. Thus, a balance that can weigh a mass to a precision of 0.01 g is a centigram balance and one that weighs to the nearest 0.001 g is a milligram balance. The *accuracy* of a measurement depends on how the mass determination is done. Sloppy technique can mean large errors in measurement.

In measuring both mass and volume (next section) it is important to protect the original materials from contamination. Therefore, dispense only the amount you need to use; if you make a mistake and take too much, then discard the excess in a proper manner.

Electronic balances are very common in the undergraduate laboratory. These work by electronically detecting the effect of a weight placed on the pan. Milligram balances are so sensitive that they can detect changes in air currents. Therefore, each balance is equipped with a draft shield that has movable windows on the sides and on the tops. If at all possible, these should be closed before recording data.

A balance must be zeroed before all measurements. To do this, close all the windows on the draft shield if present, then *gently* press the Tare button. After a moment, the display should respond with a reading of zero grams. Put your sample onto the balance and, after closing the windows on the draft shield if you can, record the mass.

When you are done, remove your sample, close all the windows, and rezero the balance. It is very tempting to use the Tare button to make the mass of a weighing vessel or paper equal to zero grams in order to simplify calculations. In some cases this is justified, but it is a bad habit to use in a common balance room. The next time someone pushes the Tare button, *your information is permanently lost and you may not even know that the value of the Tare has changed.* Therefore, always zero the balance before weighing your vessel or paper. Then record the absolute weight (to the precision of the balance) of the vessel or paper. Subtract this mass from the mass of the vessel + sample to obtain the mass of the sample.

If milligram precision is required, then fingerprints and other soil can add detectable mass to an object. Therefore, between the initial and final weighing, all containers should be handled by tongs or, if necessary, protected by a piece of laboratory tissue.

In many cases, the most precise balances will be separated in a special area or room where no transfers of chemicals should occur. In this case, good practice requires a trip to the balance room to get an accurate measure of the weight of a container or weighing paper. Then, back in the laboratory a chemical or solution is dispensed, often with a preliminary mass determination on a less precise balance. Finally, the container or paper plus the substance of solution is reweighed on the more precise balance.

There are many cases in the laboratory when you are instructed to use a mass of substance that is *close to* some particular value, for example, 2 g of metal salt. This means that, for the purpose of the experiment, there is some latitude in the mass of that substance. Any value within 10–20% of the indicated approximate mass is fine. However, it is important in most cases that you do know the precise mass of what you do use. So, even if there is latitude in the mass that you use, you must get a precise mass for your experiment. Thus, the notebook should record the mass determinations to within 0.001 or 0.01 g as indicated.

C. Measurement by Volume

Cleaning and Preparing Glassware

Many chemical procedures require the use of liquids, either pure substances or solutions. How much is used, and how precisely we determine the amount, varies widely from experiment to experiment. In some cases when we say 10 mL of water, any value between about 5 and 15 mL is fine. Then a measurement based on markings on the side of a beaker or flask is fine. But in other cases we will need to have exactly 10.00 mL of water. In that case, an instrument known as a volumetric pipet will be used.

Because solutions often contain reactive substances and because many solids dissolve readily in water, it is very important that the vessels used to handle liquids be very clean and, often, dry.

The first step in cleaning glassware is to remove any solids that are apparent on the glass. If you do not know what the solid is, then check with your instructor before cleaning the glassware. If the material is safe to dispose of in the drain, then a simple soap and water cleaning with tap water and a laboratory brush should be enough.

Rinsing with water or scrubbing with soap can leave behind a significant amount of other material. Therefore, before using most glassware it is essential that it be well rinsed with the purest water available in the lab. Usually, this will be deionized water. The rinsing does not require that the vessel be filled completely. Adding several small portions of deionized water, swirling to ensure that the water has wet the whole vessel, and then draining the vessel will be enough.

You can check for cleanliness by putting some deionized water in the vessel and then pouring the water out. Some water will stick to the inside of the vessel. This should appear as a clean sheet of liquid. If no beads form on the walls of the vessel, then it is clean and you may proceed. If beads form, wash the vessel with a soap solution and a brush, rinse with tap water, and rinse again with deionized water. Repeat until no water beads form on the inside.

In some cases you will be dispensing a chemical or a solution directly into the vessel. In that case, it is often OK if you have a few drops of pure water present in the vessel. These will not affect the mass or the number of moles of added reactant.

Dry glassware is required when the added substance or solution has to remain dry or be undiluted. In those cases, drying the clean glassware means shaking or dabbing the last drops of water from the vessel and allowing the vessel to air dry for a few minutes. Some laboratories may have an organic solvent like acetone present. This can aid drying, but acetone can interfere with many procedures and it is flammable. Always check with your instructor regarding the use of acetone in drying glassware. One tactic that should be avoided is drying using a stream of laboratory air. This is *not* expected to be clean and dry and may contaminate your clean glassware.

Some experiments involve the use of a reactant or solvent that is of a known composition or concentration. To ensure that this is not changed, after the vessel is cleaned it must be *pre-rinsed* with the liquid that it will hold. To do this, add a few milliliters of the substance or solution to the vessel. Swirl to coat the entire inside of the vessel. Then discard the substance or solution in a proper manner. Do *not* return these rinses to the original container. Discard them as instructed.

Types of Volumetric Glassware

Two factors determine what glassware should be used to measure a liquid volume: convenience and precision. These are discussed in the chart that follows, which includes the major types of glassware found in general chemistry laboratories. All of these kinds of glassware are available in a variety of sizes. The size required is determined by the amount of liquid that needs to be measured and, for beakers and Erlenmeyer flasks, whether the glassware will be used for another purpose, such as carrying out a reaction.

Using Volumetric Glassware

Burets. Most burets in use in general chemistry laboratories contain between 25 and 50 mL of solution. They are marked in 0.1-mL increments. But, as with any measured number, you can consistently get data to 0.05 or even 0.01 mL by looking very carefully at the level of the solution. The material dispensed from the buret must be the undiluted solution from the bottle you are using. To ensure this, the buret must be clean of all other material, including other solutions, and it must be pre-rinsed with the solution. The tip of the buret must be filled with the solution before the experiment is begun. Be very careful to check the tip for bubbles; these affect the accuracy of the measurement.

The flow of solution from a buret is controlled by a stopcock, which can be turned to permit the solution to leave the buret at a wide variety of rates. It is a good idea to practice with some water before using a valuable reagent. Try to get a smooth stream, a rapid flow of drops, and a flow of drops that is less than one per second. The very best control is to allow less than one full drop to form at the tip of the buret. Then, if appropriate for the experiment, a stream of deionized water from a wash bottle can be used to knock the drop off the buret tip and into the beaker or flask below.

Pipets. Two kinds of pipets are used for volumetric measurements. One, called a volumetric or transfer pipet, measures one particular volume to high accuracy. The

Type of Glassware	Issues of Convenience	Issues of Precision and Accuracy
Beakers	A wide opening at the top makes beakers easy to clean and to add material. A pouring lip makes dispensing easy.	The lines on the side of the beaker are rarely precise to more than 10%.
Erlenmeyer flasks	The narrow top is good for containing vapors and for pouring. The angle of the sides is useful for containing a stirred solution.	The lines on the side of the flask are rarely precise to better than 10%.
Burets	Burets have markings that allow for measurement of a range of volumes. The buret's stopcock can be used to dispense a steady stream of a solution or drops (even fractions of a drop).	Burets measure different volumes to a consistent precision (to 0.01 mL for a 25-mL buret)
Volumetric pipets	These allow the rapid measurement of particular volumes reproducibly	The single-volume measurement of a volumetric pipet gives the greatest precision available (for example, 0.01 mL in 25 mL)
Mohr pipets	These allow the rapid measurement of a variety of volumes.	The precision of Mohr pipets is usually greater than that of burets, sometimes 0.01 mL in a 5-mL measurement.
Droppers	Counting drops is a fast way to dispense small amounts of liquid.	The precision depends on a consistent drop size, and this requires a steady hand and constant angle. You should never assume two droppers have the same drop size.
Volumetric flask	An excellent way to quickly take a measured amount of material and use it to prepare a solution of known concentration.	Volumetric flasks are the best way to prepare an exact amount of solution of a fixed concentration. They are generally precise to ± 0.01 mL. They come in many different sizes.

other, a Mohr pipet, has gradations that permit different volumes to be dispensed very easily.

Both are controlled in the same way, through a pipet bulb that is placed atop the pipet and used to pull solution into the pipet and, sometimes, to dispense the solution. There are several different kinds of bulb. The simplest has a single opening that connects the bulb to the pipet. This is used to draw liquid into the pipet by squeezing the bulb and then placing it atop the pipet. The pipet is placed in the solution to be measured. The bulb is slowly released, and the suction draws the solution into the pipet. Be careful to keep the tip below the surface of the solution. If you do not, you may draw air very quickly into the pipet and cause the solution to shoot up into the bulb.

As the level of the solution nears the line you wish to reach, slow down on the rate by tightening the grip on the bulb. Draw a slight excess of solution into the pipet and then, without removing the bulb, lift the pipet out of the solution. With a simple bulb you should next smoothly but quickly slip a finger onto the top of the pipet in place of the bulb. Gently lifting this finger (a rolling motion may be best) allows air back into the pipet, and the solution will begin to flow out of the top. Stop when the level of the liquid is back to the desired line.

If you have a volumetric pipet, you are ready to dispense the liquid directly from the volumetric pipet. Simply lift your finger and allow the solution to drain from the pipet. A small amount of liquid may remain in the tip. *The pipet is usually designed to have that liquid remain, so do not dispense it as part of the measurement.* A Mohr pipet works by measuring the difference in a starting and ending volume. Dispense the liquid by gently lifting your finger; stop the flow by putting your finger back on the pipet.

D) Techniques of Titration

No matter what kind of chemistry you do, even if you don't pursue a career as a chemist, you are likely to have to run a **titration.** A titration occurs when we add a chemical substance to a system gradually and observe the changes. In the laboratory this usually means we use a solution with a known amount of reactant to determine the number of moles, and then the concentration, of the reactant in a second solution.

There are two reactants in any titration. The material added is called the titrant and the solution it goes into is the titrand (this latter term is rare). One solution contains a known chemical amount of a reactant. The other contains an unknown amount of a second reactant. The titrant is added to the titrand until some indication signals that the reactant in the titrand is all consumed. This is the end point. Good titration technique makes this end point very close to the volume at which a stoichiometric amount of the two reactants have been mixed and have reacted (the equivalence point).

A good titration reaction is quick and complete. The end point should be easy to detect ("sharp"). In practice, end points are easy to overshoot. In those cases, the end point is some unknown amount beyond the equivalence point. To get a more accurate determination of an end point close to the equivalence point, good technique demands that you do more than one trial of any titration. The first trial may be off, but it gives a much better idea of where the end point will be, so subsequent trials can be much more accurate. This additional accuracy comes from making the approach to the end point as gradual as possible.

There are three major types of titration reactions. One type involves acid–base reactions. These are important in determining concentrations of acids and bases. A second type involves oxidation–reduction reactions. The third type of titration is the compleximetric titration.

Dispensing the Titrant

When dispensing the titrant, it is OK to add a solution as a stream *only* if you are certain that you are not near the end point. As you approach the end point, start adding titrant a drop at a time. Often you know the end point is near because you can see, temporarily, the color change that will signal the true end point. But this color may fade as the solutions are mixed completely. The last material should be added as fractions of a drop. You do this by allowing a drop to start to form at the tip of the buret, then use a gentle stream of deionized water from a wash bottle to rinse the drop into the solution below.

The end point may be ambiguous in some cases. Therefore, record the volume of the buret and the appearance of the titration solution at several points. You may then examine the data to find the place that you think is closest to the true end point. If you overshoot, of course, you will lose information that may be important. So, *go slowly.*

Acid–Base Titrations

Most acid–base titrations involve a solution of an acid and a solution of a base. In the simplest case, the acid and the base react in a 1:1 ratio to give the corresponding conjugate base and conjugate acid:

$$acid_1 + base_2 \rightarrow base_1 + acid_2$$

This represents the Bronsted–Lowry theory of acids and bases, and is one of the clearest ways to describe how acids and bases react with each other in solution.

In all practical titrations in aqueous solution, either the starting acid or the starting base is strong. Thus, one of the reactants is either H_3O^+ or OH^-, giving water as a product:

$$H_3O^+ + B \rightarrow B\text{---}H^+ + H_2O$$

$$HA + OH^- \rightarrow A^- + H_2O$$

How do we know that an acid–base titration is complete? The simplest way is through a colored acid–base indicator. These indicators are substances that, in the presence of even a small amount of excess acid or base, rapidly change color. Therefore, we can add acid or base until the indicator changes color and then we know we've completed the titration. *For monoprotic acids and bases,* at completion, we have added an amount of acid (or base) exactly equal to the **number of moles** of the acid and base in the original solution. From this number of moles and the volume of the solutions we can determine the molarity of the unknown solution. The fundamental equation is:

$$moles\ acid = moles\ base$$

$$c_{acid}V_{acid} = c_{base}V_{base}$$

$$c_aV_a = c_bV_b$$

We can solve for any of the four quantities in this equation if we know the other three. We can also work with a known *mass* of an acid or a base and use the molar mass to find moles acid or base:

Known mass of acid with unknown [base]	**Known mass of base with unknown [acid]**
moles acid = moles base	moles acid = moles base
$\dfrac{mass\ acid}{molar\ mass\ of\ acid} = c_bV_b$	$c_aV_a = \dfrac{mass\ base}{molar\ mass\ of\ base}$

Different equations will apply in titrations where more than one H^+ is transferred.

Standards for Acid–Base Titrations

Two kinds of standards are used in analytical chemistry. The first type, called a *primary* standard, is a highly purified compound that is weighed accurately on a balance before it is dissolved and used in the titration. A *secondary* standard is one whose concentration is determined from a primary standard and then used in other reactions. Secondary standards are usually less expensive and easier to handle than primary standards. However, most secondary standards will change in concentration with time. For example, strong bases will react with carbon dioxide in the air.

E. Dilution Calculations

In many experiments and procedures, volumetric glassware is used to prepare solutions from more concentrated solutions known as stock solutions. Here is a summary of the concepts and steps involved.

When we dilute a solution, we increase the volume without changing the chemical amount (commonly, moles) of the substances that are dissolved in the solution. This means that, although the volume V and the concentration c change, the number of moles do not. If we express concentration in units of amount per volume, then concentration times volume, $c \times V$, is equal to the amount. If we are tracking moles:

$$n_{\text{initial}} = n_{\text{final}}$$

$$c_{\text{initial}}V_{\text{initial}} = c_{\text{final}}V_{\text{final}}$$

It is common to abbreviate final as f and initial as i, so this equation becomes $c_iV_i = c_fV_f$. Let's use this in some sample problems

Dilution of a Known Concentration by a Known Amount

If we dilute 2.50 mL of 0.100 M Cu^{2+} to 25.00 mL in a volumetric flask, what is the final concentration of Cu(II) ion?

Answer. In this case c_i is 0.100 mol L^{-1}, V_i is 2.50 mL, and V_f is 25.00 mL. We solve for $c_f = c_iV_i/ V_f = (0.100$ M$)(2.50$ mL$/25.00$ mL$) = 0.0100$ M. Note that we do not have to change the volume to liters, because both volumes have the same measurement, milliliters.

Determining How Much to Dilute a Solution

We need to prepare 0.500 L of a solution of 0.0250 M Fe^{3+}. We have a stock solution of 0.528 M Fe^{3+}. How should we prepare the solution?

Answer. Here we are seeking the value for V_i. This will be

$$V_i = (c_f/c_i)V_f = (0.0250 \text{ M}/0.528 \text{ M}) (0.500 \text{ L}) = 0.0237 \text{ L}$$

In this case, we will take 0.0237 L, or 23.7 mL, of the stock solution and place it in a 0.500-L volumetric flask. Addition of water, with mixing, to give a 0.500 L total will give the required solution. Note that it is incorrect to say we will add 472.3 mL of water, because that assumes that volumes are additive. This is an assumption we must make sometimes, but never if we can use volumetric glassware with precise volume amounts.

Finding a Concentration When Solutions Are Mixed

There are times that the most efficient, though not the most precise, way to make a solution is by mixing simple volumes. For example, if we mix 3.80 mL of 0.50 M NH_3, 0.30 mL of 0.100 M Ag^+, and 0.90 mL of water, what is the concentration of the $[Ag(NH_3)_2^+]$ that will form, assuming that volumes are additive and that all of the silver ion becomes $[Ag(NH_3)_2^+]$?

Answer.   The total volume of this mixture will be, approximately, 5.00 mL. We note that the number of moles of $[Ag(NH_3)_2^+]$ will equal the number of moles of Ag^+, so we can calculate the concentration of $[Ag(NH_3)_2^+]$ from the concentration of Ag^+ that we add:

$$c_f = c_i V_i / V_f = (0.100 \text{ M})(0.30 \text{ mL}/5.00 \text{ mL}) = 0.0060 \text{ M}$$

IV. PRINCIPLES OF SPECTROPHOTOMETRIC ANALYSIS

A. The Measurement of Color

The human eye and brain are extremely sensitive to color. When the eye and brain sense a spectrum of visible light that is the same as that of the sun, then they perceive either white or a shade of gray. If the eye and brain detect only a section of the spectrum, then a color is perceived.

There are two ways an object can come to have color. First, the object can be a source of light of a restricted wavelength. Under these circumstances, the observer perceives the wavelength that comes from the emission of light by the object. Think of the bright yellow lights that are often used for exterior lighting. These use the emission of light by sodium vapor to give a bright light that is tightly restricted to the region at about 590 nm (one nanometer is 1×10^{-9} m), at the center of the yellow region of the spectrum.*

Color can also arise when an object absorbs some wavelengths but reflects or transmits others. This selectivity arises from the absorption of light by the object. An observer perceives the *complement* of the light that is absorbed. A color wheel (Figure I-1) helps us to understand and predict the appearance of a simple absorption. For example, when light of blue wavelengths is absorbed, then light of violet, green, yellow, orange, and red wavelengths is transmitted or reflected. This combination appears to our eyes as orange.

The eye and brain have limitations in ascertaining which wavelengths of light are absorbed by a sample. For example, say you see the color red. This can arise because only green light is being absorbed by a sample, in which case your brain combines the perception of transmitted red, orange, yellow, blue, and violet light as red. On the other hand, the color red is perceived in cases where *all* wavelengths of light

FIGURE I-1   A COLOR WHEEL TO DETERMINE COMPLIMENTARY COLORS

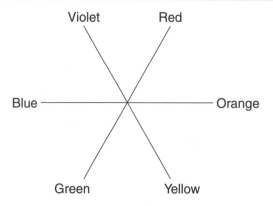

---

* You can tell that this light is of only a narrow wavelength range because, if you look at a color picture illuminated by such a sodium vapor light the picture appears sepia!

except red are absorbed and red light wavelengths are reflected or transmitted. Your senses cannot tell the difference.

Another characteristic of the eye and brain is that they are more sensitive to certain wavelengths and less sensitive to others. The peak sensitivity is in the region where the sunlight is concentrated: around 500 nm. This is the green–yellow region of the spectrum. The great sensitivity of the human visual system to green–yellow light is the reason this color is often used on emergency vehicles and taxicabs.

Scientists study light and color quantitatively through the use of spectrophotometry—the quantitative determination of light intensity by wavelength. The signal from an electronic detector changes depending on the intensity of light striking the detector. These detectors are (generally) equally sensitive over a wide range of light wavelengths. To determine the amount of light at different wavelengths, one must separate the light into its components. The standard way to do this is through the use of a prism or diffraction grating that bends beams of different wavelengths in different amounts. The gathering of information concerning intensity vs. wavelength is called *spectrophotometry*, and the corresponding instruments are called *spectrophotometers*.

Spectrophotometry can be performed across the electromagnetic spectrum (Figure I-2), from gamma rays (useful in nuclear chemistry) to radio waves (used in astronomy). Spectrophotometers designed for wavelengths in the ultraviolet (180–380 nm) and visible (380–700 nm) regions are common. They are referred to as UV–visible spectrophotometers.

B. Absorbance and Transmittance

There are several different ways of quantifying the amount of light that comes through a sample. The simplest to understand is percent transmission (%T), which is the percentage of light that comes through the sample at a given wavelength. A far more meaningful number in practice is the absorbance, A. This is equal to a logarithmic function of %T:

$$A = -\log_{10}\left(\frac{\%T}{100}\right)$$

Absorbance is an open-ended scale and has no units. When all of the light gets through (%T = 100), then A = 0. When %T = 10%, A = 1.0. When %T = 1%, then A = 2.0. When no light is transmitted (%T = 0) then A = ∞.

The relationship between absorbance and concentration can be a simple one:

$$A = \varepsilon c l$$

FIGURE I-2 **THE ELECTROMAGNETIC SPECTRUM**

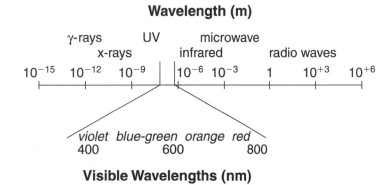

Visible Wavelengths (nm)

where c equals concentration in moles liter^{-1} (mol L^{-1}), l equals path length (typically 1 cm), and ε equals the *extinction coefficient*, which in these systems has units of L mol^{-1} cm^{-1}. This relationship is known as *Beer's Law*, and it is the basis for a great deal of spectrophotometric analysis. Beer's Law is reliable only when A is small—say under 2—and it relies on the assumption that the nature of the substance under study does not change with concentration.

The extinction coefficient indicates how much of the light that strikes a sample is absorbed by a given concentration of the chemical substance. Small extinction coefficients are in the range 1 to 1000 L mol^{-1} cm^{-1}. Large extinction coefficients can range up to 100,000 or more. You can also think of the extinction coefficient as an efficiency indicator, and indeed systems that need to operate at high efficiency (such as the photosynthetic apparatus in plants) have very large extinction coefficients. Food colorings, which give a lot of color for very little substance, also have large extinction coefficients.

C. Deviations from Beer's Law

Beer's Law deceives us with its simplicity, because it suggests that a linear relationship exists between absorbance and concentration. This relationship holds for many, but not all, substances. Deviations can occur for at least three reasons:

1. The molecules that absorb light interact with one another instead of absorbing light independently.
2. The light-absorbing substance forms incompletely from the added reagents because of equilibrium or side reactions.
3. The light-absorbing substance forms slowly compared to the time required to make an absorbance measurement.

In all three cases, we can find out whether Beer's Law applies by testing a series of different solutions prepared with different amounts of reagents. This is used to construct a graph where the y-axis has units of absorbance and the x-axis has units of concentration. Such a graph is called a calibration curve. If the graph of A vs. c is a line, then Beer's Law applies. If it is not a line, then we can still determine c for an unknown solution by comparing A with the curve that the standard solutions provide.

It is also important to observe, and possibly remeasure, solutions after they have been allowed to stand, to check if the absorbance depends on time.

D. Spectrophotometric Samples

Most spectrophotometers use small glass or plastic cuvettes to hold the sample. They have a square design, with two clear sides for the light to shine through. A 1-cm cuvette typically holds 3 mL of solution.

The most accurate technique requires precise control of solution concentrations, so volumetric glassware, including volumetric flasks, is used to prepare samples. Then the sample is transferred to the cuvette to fill it.

It is also possible to get good results by preparing the test sample right in the cuvette, saving glassware and time. If this is done, you must remember three points:

1. All samples must have the same volume. If you mix 2.00 mL of a reagent with 0.30 mL of a second reagent, don't forget to add enough of the correct standard to make the volume the same total volume in each case.
2. Preparing a solution by mixing measured amounts of other solutions assumes that the volumes are additive. In these labs, we have verified that it is a reasonable assumption, but in other situations be aware that the additivity of solution volumes is not valid.

3. The volume of a cuvette is small. Therefore, plan to work with small amounts of reagents, and do not vary the amount of the "main" reagent. This can be done by using 2.0 mL of the main reagent in every spectrophotometric sample.

E. Using the Spectronic-20

The Spectronic-20 spectrophotometer (Figure I-3) uses a tungsten lamp as its source of light for the analysis of compounds that absorb in the UV–visible range ($\lambda = 350$–800 nm). The optical components of the instrument itself focus the incoming light with mirrors and direct it through a process that allows one to select the particular wavelength of light that finally passes through the sample. In the Spec-20, light from the source passes through an entrance slit and is focused by a concave mirror onto a reflection grating. The grating serves to disperse the light according to the various wavelengths present. Particular wavelengths can be selected by rotating the reflection grating by means of an outside knob that allows you to "dial in" the wavelength you desire. The selected wavelengths of light reflect off the grating and are focused by a second mirror to pass through a sample. Light emerging from the sample strikes a detector that analyzes its intensity. A comparison of the amount of light entering the sample with the amount of light leaving the sample gives a measurement of the amount of light that was absorbed by the sample.

Several techniques must be understood and mastered before you attempt to use any spectrophotometer. Good results depend on a measurement of the light that is absorbed by a solution of the sample. The outside of the cuvette or container holding the sample must be clean. Wipe the surface so that it is free of fingerprints and moisture. Some cuvettes have two frosted and two clear sides. These must be inserted into the sample compartment so that the light passes through the clear sides. The lid to the sample compartment must be closed before taking a measurement so that all of the selected light passes through the sample. The instrument must be adjusted to correctly register both zero and one hundred percent transmission before each use. The procedure outlined here is a good one to follow.

1. Allow about 20 min for the instrument to warm up.
2. Adjust the wavelength control knob to the desired wavelength.
3. Prepare the instrument to make accurate measurements for your particular solution by making the following adjustments:

   a. With no sample in the sample compartment, adjust the dark current control to read zero percent transmission ($\%T$).
   b. With a blank solution in the sample compartment, adjust the light control knob to read 100% transmission. The blank solution may be distilled water,

FIGURE I-3 THE CONTROLS OF THE SPECTRONIC-20 SPECTROPHOTOMETER

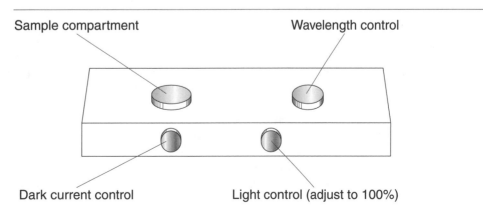

Sample compartment   Wavelength control

Dark current control   Light control (adjust to 100%)

or it may be the solution that will eventually contain your sample. Specific directions for making up a blank solution are generally given within a laboratory procedure.

4. With these steps done, you are ready to measure the absorbance of your sample. Insert the cuvette containing your sample into the sample compartment, close the lid, and record both the absorbance and the % transmission (%T) readings. The scale for %T may be easier to read accurately, and it can readily be converted into absorbance through the relationship, $A = -\log(\%T/100)$.

F. Using a Recording Spectrophotometer

Some spectrophotometers can record the absorbance of a spectrum at many different wavelengths. Those in common use have an open design, which means that the sample chamber is directly accessible. Calibration is done by recording a blank spectrum. This is then used for comparison with sample spectra. The optical system of the most common laboratory spectrophotometer includes a tightly focused beam of light that passes through the sample and is then directed onto a grating. This grating, which does not move, disperses the light onto an array of diodes. The dispersal and the arrangement of the diodes permit the division of the light into 2-nm-wide channels. Each is detected separately and simultaneously by the instrument. Thus, the spectrophotometer functions similarly to the Spectronic-20, except that the use of the array and computer-controlled data sampling enables the instrument to monitor hundreds of wavelengths at a time. A limitation of this design is that the resolution can never be better than 2 nm. A scanning spectrophotometer (whether automated or manual) can select wavelengths with much greater precision.

G. Calibration Curves

Many spectrophotometric procedures require you to prepare a calibration curve for analysis of unknowns. A calibration curve is *your* ruler to measure a chemical system. This must be done well, because all later data will be referenced to this one set of measurements.

You start the process of calibration with an accurately determined concentration of a substance in a stock solution. This is then diluted to give a set of solutions that span the range of possible concentrations in the experiment. Sometimes these standards are placed right in cuvettes. If necessary, these standard samples are combined with other reagents to give the sample for the instrument.

In Figure I-4 we present data for the calibration curve one might use in determining the concentration of the dye carmine indigo. It does not need a second reagent to give a suitable spectrophotometric sample. We take a sample of a solution of carmine indigo and place it in a cuvette with no further manipulation. The range of possible values we must "cover" with the curve is equal to the range of values we may have in the actual sample.

The figure shows the data in three ways. First, notebook entries indicate the volume of stock solution mixed with water to prepare the standard solution. Note that the notebook indicates exactly how the sample was prepared. Second, the solutions were transferred to a spectrophotometer and the absorption spectrum was measured, giving A values for our notebook (again, record *all* essential data in the notebook). The acceptable range of values of A for spectrophotometry is generally 0.5–1.5. We can see that the data are in the acceptable range of values, so we can use this method to prepare a calibration curve.

These data were transformed into the graph at the bottom of the figure. Note that the graph has five data points, one for each standard, and a line connecting them. Dotted lines are used to extend the curve to the end of the graph region. These are

Notebook records

All solutions prepared by diluting
stock solution with deionized water
to make 100ml solution.

mL stock	[dye]	A(600nm)
2.00	1.72×10^{-5}	0.204
4.00	3.43×10^{-5}	0.421
6.00	5.15×10^{-5}	0.652
8.00	6.86×10^{-5}	0.890
10.00	8.59×10^{-5}	1.119

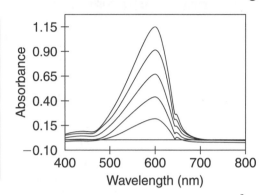

Calibration Data: Carmine Indigo

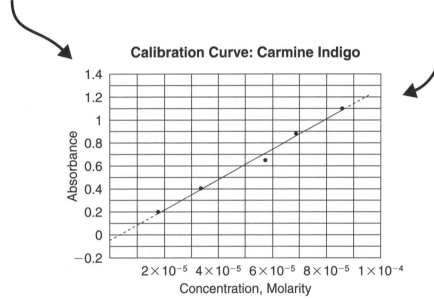

Calibration Curve: Carmine Indigo

FIGURE I-4 **SCHEMATIC PREPARATION OF A CALIBRATION CURVE**

extrapolated lines, because we did not actually measure these regions. One should only use an extrapolation of a graph in emergencies.

The data analysis suggested in the figure should be done *during the lab period*, including a preliminary calibration curve. If any data points seem to lie off the curve of the other data, then there is ample time to remeasure those particular points.

In the second kind of calibration curve, we mix solutions of a substance with another reagent, yielding a colored substance for measurement. We prepare the standards, as before, to cover the range of possible concentrations in our unknown solution. All of these—standards and unknowns—are then mixed to prepare solutions for the spectrophotometer. This is often done right in the cuvette, although very accurate work requires that this step be done in a volumetric flask. We must mix a consistent volume of our standards and unknowns with a consistent amount of the solutions that contain the additional reagent or reagents. These are then examined with the spectrophotometer. After verifying that the data are consistent with a good spectrophotometric calibration experiment, we plot a calibration curve, and this allows us to read the curve to find how much of a substance is in the unknown.

V. QUANTITATIVE INTERPRETATION OF CHEMICAL MEASUREMENTS

A. Chemical Reaction Stoichiometry

The measurement of the amounts of substances involved in a chemical reaction is known as chemical stoichiometry. It is an essential part of most general chemistry courses, and here we review only a few general points.

Measurement of Chemical Amount

Chemists cannot measure the number of molecules in a sample directly, except under very unusual circumstances. Instead, they measure the amount by other means, especially by measuring mass. We connect a count of molecules or formula units with mass through the formula mass, which is determined by multiplying the atomic mass of each element present by the number of times its symbol appears in the formula. Atomic and formula mass are linked to measurable masses in the laboratory. For chemists, this mass is the *gram*. We define the mass of one mole of a substance as equal to the formula mass, expressed in grams. Molar mass is the mass, in grams, of one mole of a substance.

For example, one mole of carbon dioxide, CO_2, has a mass of 12.011 g + 2(15.9994 g) = 44.0098 g (calculator answer) or 44.010 g (rounded answer):

$$\text{molar mass } CO_2 = 44.010 \text{ g mol}^{-1}$$

The molar mass of a chemical substance is one of its most important properties. Why? Because we cannot directly measure the number of formula units or molecules in a sample; there are no "molemeters" in the laboratory. Instead we can easily measure the mass of a sample. And from the mass of a substance and its molar mass, we can determine the number of moles in a sample. The molar mass is the constant of proportionality relating mass and moles.

Molar mass can also be used to tell us how many grams of substance give a certain number of moles. This is critical if chemists are to measure the right amount of substance. For example, if we want to know the number of moles of iron in a sample of iron that has a mass of 223.22 g we can use the molar mass of iron to carry out a unit conversion calculation:

$$223.22 \text{ g Fe} \times \frac{1 \text{ mol Fe}}{55.847 \text{ g Fe}} = 3.9970 \text{ mol Fe}$$

Chemical Amounts in Reactions

Chemical substances can react with one another to form new substances. This is called a chemical reaction and is shown in symbols by a chemical equation. In a chemical equation, the substances that react (called reactants) appear to the left of an arrow and the new substances that are produced (called products) appear to the right of the arrow.

$$\text{reactants} \rightarrow \text{products}$$

In addition to showing reactant and product substances, a correctly written chemical equation shows the ratio these substances have to one another. Not all substances react in a one-to-one ratio. The ratio of substances in a chemical equation is given by coefficients, the numbers written in front of the chemical formula representing the substance. For example, the chemical equation for the reaction of hydrogen gas and nitrogen gas to form ammonia gas is,

$$3 \text{ H}_2 + \text{N}_2 \rightarrow 2 \text{ NH}_3$$

The coefficients in a balanced equation show how many molecules react and are produced in the chemical reaction. Thus, this equation means "1 molecule of N_2 + 3 molecules of H_2 react to form 2 molecules of NH_3." Because the mole is a group counting number that can be used to count large numbers of atoms and molecules, we can also say the equation means "1 mol of N_2 + 3 mol of H_2 react to form 2 mol of NH_3." We can then use this to determine how many moles of ammonia will be produced from the reaction of 0.50 mol of hydrogen and an excess of nitrogen according to the balanced equation. We can use the ratio

$$\frac{2 \text{ mol NH}_3}{3 \text{ mol H}_2}$$

as a conversion factor between moles of H_2 and moles of NH_3

$$0.50 \text{ mol H}_2 \times \frac{2 \text{ mol NH}_3}{3 \text{ mol H}_2} = 0.33 \text{ mol NH}_3$$

Many of the calculations in the laboratory actually require that we begin with the mass of one substance and end with the calculation of the mass of another substance. This means we must add a mass-to-mole calculation to the beginning of the process and a mole to mass calculation to the end.

For example, iron metal reacts with oxygen (is oxidized) to form iron(III) oxide according to the reaction $4 \text{ Fe}(s) + 3 \text{ O}_2(g) \rightarrow 2 \text{ Fe}_2\text{O}_3(s)$. We can use stoichiometry to determine how many grams of iron are needed to produce 100.0 g of iron(III) oxide. It is helpful to start with a "calculation map":

$$100.0 \text{ grams Fe}_2\text{O}_3 \rightarrow \text{moles Fe}_2\text{O}_3 \rightarrow \text{moles Fe} \rightarrow \text{grams Fe}$$

$$100.0 \text{ g Fe}_2\text{O}_3 \times \frac{1 \text{ mol Fe}_2\text{O}_3}{159.691 \text{ g}} \times \frac{4 \text{ mol Fe}}{2 \text{ mol Fe}_2\text{O}_3} \times \frac{55.847 \text{ g Fe}}{1 \text{ mol Fe}} = 69.9438 \approx 69.94 \text{ g Fe}$$

Limiting Reactants

The coefficients of a balanced equation are the exact number of moles of reactants needed as well as the exact number of moles of products formed. The exact number of moles is called the stoichiometric amount. When you do *not* have exact stoichiometric amounts of reactants, one of the reactants will be used up *before* the others. This will stop the reaction. The reactant that is used up is called the **limiting reactant** or limiting reagent (L.R.) because it is the substance that "limits" the amount of product that can be formed.

Determination of a limiting reactant in a chemical reaction is not always so obvious. One method used to solve limiting reactant problems is to determine the amount of product that will result using *each of the given quantities one at a time. The reactant that produces the smallest amount of the same product is the limiting reactant.* When the L.R. is gone, the reaction stops. All given masses for reactants that actually are in excess (reactants that are *not* the L.R.) are meaningless in terms of determining the amount of product formed.

For example, copper(II) sulfate and barium chloride solutions react to produce a precipitate of barium sulfate. (A precipitate is a solid that appears "suddenly" or "precipitously" when two solutions are mixed.) The reaction occurs according to the equation:

$$CuSO_4 \ (aq) + BaCl_2 \ (aq) \rightarrow CuCl_2 \ (aq) + BaSO_4 \ (s)$$

If we have 5.255 g of copper(II) sulfate and 8.230 g of barium chloride reacting, then to calculate the expected mass of $BaSO_4$, we need to do two calculations. The first calculation determines the mass of barium sulfate produced when you start with 5.255 g of copper(II) sulfate. The second calculation determines the mass of barium sulfate produced when you start with 8.230 g of barium chloride. The calculation resulting in the smaller amount of product is the one using the limiting reactant. The smaller amount of product is the only correct answer.

$$5.255 \ \text{g CuSO}_4 \times \frac{1 \ \text{mol CuSO}_4}{159.608 \ \text{g CuSO}_4} \times \frac{1 \ \text{mol BaSO}_4}{1 \ \text{mol CuSO}_4} \times \frac{233.392 \ \text{g BaSO}_4}{1 \ \text{mol BaSO}_4} =$$

$$7.6843 \ \text{g BaSO}_4$$

$$8.230 \ \text{g BaCl}_2 \times \frac{1 \ \text{mol BaCl}_2}{208.236 \ \text{g BaCl}_2} \times \frac{1 \ \text{mol BaSO}_4}{1 \ \text{mol BaCl}_2} \times \frac{233.392 \ \text{g BaSO}_4}{1 \ \text{mol BaSO}_4} =$$

$$9.2242 \ \text{g BaSO}_4$$

The first calculation produces the smaller amount of $BaSO_4$. This means that $CuSO_4$ is the limiting reactant. The amount of $CuSO_4$ present at the beginning of the reaction limits the amount of product to only 7.684 g $BaSO_4$. The expected mass of precipitate is, therefore, 7.684 g $BaSO_4$. The other reactant, $BaCl_2$ is in excess.

Reaction Yield

The amount of product actually obtained in a reaction is called the *actual yield* of the reaction. The amount of product predicted by stoichiometry is called the *theoretical yield*. The actual yield should never be greater than the theoretical yield, but it can be less—much less in some cases. The theoretical yield and the actual yield form a ratio referred to as the *percentage yield*, where

$$\text{percentage yield} = \frac{\text{actual yield}}{\text{theoretical yield}} \times 100\%$$

As an example, the synthesis of silver chloride involves mixing known amounts of a chloride salt—say, potassium chloride—and silver nitrate, a soluble silver salt:

$$KCl \ (aq) + AgNO_3 \ (aq) \rightarrow AgCl \ (s) + KCl \ (s)$$

If we mix one mole of KCl and one mole of $AgNO_3$, we should get one mole of AgCl. This would have a mass of 143.32 g of AgCl. This mass is the theoretical yield. In a typical reaction, though, only 125 g of AgCl may be isolated. This is the actual yield. The percentage yield is then:

$$\frac{125 \ \text{g AgCl}}{143.22 \ \text{g AgCl}} \times 100\% = 87.3\%$$

B. Uncertainty Analysis

Chemistry experiments usually require measurement of one or more parameters that describe the particular chemical system studied. Most laboratory measurements are very common ones such as mass, temperature, volume, or pH. It is important that you assess the reliability or trustworthiness of your measurements before you draw conclusions based on your experimental data.

Experimental results are judged using two criteria: accuracy and precision. The **accuracy** of a measurement tells us how closely it agrees with the true value of that parameter. Chemical properties such as density have known values. For these properties, the difference between our measured value and the "true" value is an indication of the measurement's accuracy. The **precision** of the measurement indicates how closely each measurement agrees with the other measurements taken for the same parameter.

Accuracy in Measurement

Several types of error affect your results: systematic error, random error, and personal error. **Personal errors** such as spillage and careless measurements must be noted whenever they occur. With practice, such errors will decrease and not greatly affect your experimental results. Personal errors affect both the accuracy and the precision of your results. The most common kind of personal error occurs when something is done too quickly. For all measurements, take your time and, if possible, repeat the measurement. When transferring material, be sure that you transfer everything that can be properly delivered by a measuring device.

Systematic error occurs when there is a flaw in the measuring device that results in measurements that always occur in the same direction (too high or too low). Systematic error affects the accuracy of your result. An example of systematic error is an improperly calibrated thermometer or balance. When we are aware of a systematic error, we try to correct it. In general, however, we are usually unaware of the error built into a particular measuring device. There are occasions when we seek the *difference* between two measurements (mass gained or lost, temperature change, etc.). In those cases, systematic errors may be equal to each other and can be ignored.

Random error occurs because of small, uncontrollable variables in one's ability to obtain exactly the same numerical value for numerous measurements of the same quantity. Random error can be decreased but not entirely eliminated. Successive measurements of the same property result in approximately but not exactly the same numerical value. Measurements cluster about an **average value** or **mean,** with some measurements slightly higher and some slightly lower than the mean value. The difference between a particular measurement and the mean value is called its deviation from the mean. The degree of deviation from the mean is used to assess the precision of experimental results. Slight deviations from the mean represent good experimental precision; large deviations indicate poor precision.

Precision in Measurement

The precision of experimental results is calculated using the average deviation from the mean value. The mean is found by summing the individual measurements (X_i) and dividing by the total number of measurements summed (N).

$$\text{mean} = \overline{X} = \frac{\sum_{i=1}^{N} X_i}{N}$$

When only two or three results are available, we must rely on the range of values from lowest to highest as an indication of experimental precision. For example, suppose that you determine the density of an unknown metal as 2.978, 2.885, and 2.994 g/cm^3. The average value is

$$\frac{2.978 + 2.885 + 2.994}{3} = 2.946 \ g/cm^3$$

The precision of these measurements is shown in the range of values reported; from the lowest, 2.885 g/cm^3 to the highest, 2.994 g/cm^3, there is a difference of $2.994 - 2.885 = 0.109 \ g/cm^3$. Thus all reported values fall within an average deviation of

$0.109/2 = \pm 0.054$ g/cm^3 from the mean and the results are reported as 2.946 ± 0.054 g/cm^3. The $\pm$ notation indicates the precision of the experimental results as it tells how closely the values are clustered (within the bounds of 0.054 higher and lower than 2.946).

Statistical analysis is used to determine the precision of experimental results for larger bodies of data, ideally 20 or more measurements although it is often used whenever $N > 3$. A statistical treatment of random error gives the standard deviation as a measure of experimental precision. The **standard deviation** (s.d.) is the square root of the sum of the squares of each measurement's deviation from the mean divided by $(N - 1)$.

$$\text{s.d.} = \sqrt{\frac{\sum_{i=1}^{N} (X_i - \overline{X})^2}{N - 1}}$$

To calculate the standard deviation, find the mean value $\overline{X}$, find each deviation from the mean $(X_i - \overline{X})$, square each deviation, sum the squares, divide by $N - 1$, and finally take the square root of that number.

Example. Find the standard deviation for these experimental results: 1.1244, 1.1198, 1.1232, 1.1190, 1.1204 g/L.

Measurements, g/L	$(X_i - \overline{X})$	$(X_i - \overline{X})^2$
1.1244	0.00304	0.00000924
1.1198	−0.00156	0.00000243
1.1232	0.00184	0.00000338
1.1190	−0.00236	0.00000557
1.1204	−0.00096	0.000000922

$\overline{X} = 1.1213_6$*

$$\sum_{i=1}^{N} (X_i - \overline{X})^2 = 0.0000215$$

$$\frac{\sum_{i=1}^{N} (X_i - \overline{X})^2}{N - 1} = 0.00005375$$

$$\text{s.d.} = \sqrt{\frac{\sum_{i=1}^{N} (X_i - \overline{X})^2}{N - 1}} = 0.00232$$

The standard deviation is 0.0023_2 and is rounded to the same number of decimal places as the mean. The result is reported as mean $\pm$ s.d.; 1.1214 ± 0.0023 g/L.

Uncertainty Analysis and Propagation of Error

There is some uncertainty in all measurements we do in the laboratory. **Uncertainty analysis** estimates the error introduced into a measurement due to random error. The instruments and glassware we use in the lab vary in their abilities to give precise measurements. The range of their precision is indicated by a $\pm$ symbol followed by numbers that indicate the degree of precision for that measurement. For example, a mass of 50.00 ± 0.01 g indicates a balance that is precise to ± 0.01 g in any mass measurement; the "actual" mass lies between $50.00 + 0.01$ g and $50.00 - 0.01$ g. We

---

* The subscripted digit is the first insignificant digit. It is carried through the calculation and rounded off at the end.

can be certain only that the substance has a mass somewhere between 49.99 and 50.01 g. A mass of 50.000 ± 0.001 g indicates a balance precise to ±0.001 g. The second balance is more precise than the first balance because it results in masses precise to thousandths rather than hundredths of a gram.

All directly measured parameters in an experiment have uncertainties. When these measurements are used to calculate some final result, their individual uncertainties will contribute to an overall uncertainty in the final calculated result. *Uncertainty is propagated in a calculation using directly measured parameters that contain uncertainties.* The final, overall uncertainty in a calculated result is obtained by doing three calculations. Using the initial uncertainties, we calculate the lowest possible value, the central value, and the highest possible value. Rules for these calculations follow.

- The symbol ± gives a high and a low limit for each measurement. This results in three values for each measurement: a low limit value, a central value (the measurement itself), and a high limit value.
- For addition and multiplication, add or multiply limits on the same side (high limit with high limit; low limit with low limit).
- For subtraction and division, subtract or divide limits on the opposite side. In subtraction problems, the high limit results when the minuend (top number in a subtraction problem) is the high limit value for that measurement and the subtrahend (bottom number in problem) is the low limit value for the measurement. In division problems, the high limit results when the numerator is the high limit value and the denominator is the low limit value for the appropriate measurements.

As an example, suppose you perform an experiment to measure the density of an unknown solid substance by water displacement. You measure the mass of your sample as 8.846 g on a balance that has an uncertainty of ±0.001 g.

	Low-Side Limit	Central Value	High-Side Limit
Mass of solid (g)	8.845	8.846	8.847

The volume of the solid is the difference in water volumes (with and without the solid) using a 100-mL graduated cylinder that has an uncertainty of ±0.2 mL.

	Low-Side Limit	Central Value	High-Side Limit
Volume of water (mL)	24.6	24.8	25.0
Volume of water + solid (mL)	27.4	27.6	27.8

To obtain density, we must first determine the volume of water displaced, then divide mass by this volume. The volume of the solid is determined by subtraction:

(volume of water + solid) − (volume of water) = volume of solid

Using the rules for subtraction, opposite side limits are used. This results in the following volumes.

Low-side limit: 27.4 − 25.0 = 2.4 mL
Middle value: 27.6 − 24.8 = 2.8 mL
High-side limit: 27.8 − 24.6 = 3.2 mL

The three sets of values needed for a calculation of density are

	Low-Side Limit	Central Value	High-Side Limit
Mass of solid (g)	8.845	8.846	8.847
Volume of solid (mL)	2.4	2.8	3.2

The density calculation requires division, so again opposite sides are combined to get the high- and the low-side limits for the density.

Low-side limit: $\dfrac{8.845 \text{ g}}{3.2 \text{ mL}} = 2.76 \approx 2.8$ g/mL

Middle value: $\dfrac{8.846 \text{ g}}{2.8 \text{ mL}} = 3.159 \approx 3.2$ g/mL

High-side limit: $\dfrac{8.847 \text{ g}}{2.4 \text{ mL}} = 3.68 \approx 3.7$ g/mL

The high and low limits give the uncertainty in the end result: density = 3.2 ± 0.4 g/mL. The calculated density value contains approximately 10% uncertainty as a result of the initial instrumental uncertainties carried or propagated through the subtraction and division processes. The final percent rounds to one significant figure in this case.

$$\frac{0.4}{3.2} \times 100 = 12.5 \approx 10\%$$

C. Graphing Laboratory Data

Principles of Good Graphing

The results of many experiments in the laboratory require the use of a graph for proper interpretation. Constructing a good graph can make a huge difference in the quality of a lab analysis, so it is worth the small amount of planning that good graphing involves.

As an example, look at Figure I-4. This figure shows how a notebook results, instrument readings, and chemical calculations are used to construct a calibration curve. The principles that went into the graph are

- **Proper Identification of Dependent and Independent Variables.** In most experiments, we can identify a variable that is changed deliberately, and we call this the independent variable. The values of the independent variable are plotted on the x-axis of the graph. Another variable will be measured to see how it changes in response to the independent variable. This is the *dependent* variable, and its values are plotted on the y-axis. In the case of Figure I-4, a calibration curve describes how the amount of light absorbed by a sample (absorbance, abbreviated A) *depends* on the concentration of a dye (in this case, in units of moles per liter, or molarity). In the notebook records we see that the student actually recorded the volume of stock solution of dye used in each sample. She then *calculated* the concentration, which is the independent variable. The absorbance is the dependent variable.
- **Proper Labeling of Graph.** A graph presents a large amount of information, sometimes even an entire day's work, in one compact form. A *title* should indicate the nature of the graph (calibration curve), the system studied (carmine indigo), and, sometimes, the origin of the data (date and name). The *axis label* should spell out the variable and give the units. The axes should also indicate

the values of the variables. These values may be integers, decimal, or scientific notation.

- **Scales.** A common error in graphing is the use of inappropriate scales for the axes. Two issues are important: the inclusion of an origin point and proper use of available space. In some experiments, a point where the values of both variables are zero should be included. This is the case with the calibration curve, where we expect that the absorbance will be zero when the concentration is zero. Take a moment to look at the available graph paper. What is the largest value you have for a variable? What is the smallest value (it may not always be zero). How does the range of values relate to the number of major grid points on the paper? Try to label and scale the axes so they fit the grid neatly while using as much of the paper as possible. The grid lines do *not* have to be integers, or even decimals. Scientific notation can be used, as is done with the *x*-axis (concentration) in Figure I-4.

- **Data Points.** The data points that relate the independent and the dependent variable form an *ordered pair* of data. Each ordered pair should be marked clearly on the graph by an obvious dot or square.

- **The Curve or Line.** The data points are the measured data for a graph, but sometimes, as with a calibration curve, we are more interested in what these indicate about other values that we have not measured. If that is the case, it is important to graph a good curve or line that shows how the data points relate to each other. It is generally poor practice to simply "connect the dots" in the graph. Instead, a smooth curve or line that fits all the data should be drawn. When the line is *between* measured points, then it should be a solid line. When the line extends *beyond* the measured data (including to the origin), then it should be a *dotted* line.

Interpreting Graphical Data

Examining a graph can reveal the relationship between the dependent and the independent variables, or allow a known relationship to be interpreted. But before the data are interpreted the graph should be examined to determine if the data themselves are reasonable. In beginning science courses you are likely to see data that have a smooth relationship between the variables. If one or more data points lie well away from the line or curve indicated by the others, then you may have an erroneous data point. That is why it is essential in many experiments to draw out a preliminary graph during the lab, when such data can be recollected.

In interpreting the graph we seek to find a relationship that suggests something that physically or chemically connects the variables. There are three kinds of relationships you are likely to find between dependent and independent variables.

The value of the dependent variable may not change as the value of the independent variable changes. The graph appears "flat" as we move from left to right. There is *no dependence* in this case.

If the data points lie on or close to a line, then the value of the dependent variable has a *linear dependence* on the independent variable. The simplest case of a linear dependence is a simple proportionality: When we double the value of the independent variable, the value of the dependent variable also doubles. In that case, we expect that the origin (0, 0) will also lie on the line relating the variables. In other cases, the data may lie on a line with a formula $y = mx + b$, where m is the slope and b is the *y*-intercept of the line (where the line crosses the *y*-axis).

When linear dependence is present, we can use the data to get a line by comparing the values of two data points or two points on the line that we draw. In that case, we use the formula

$$m = \frac{y_2 - y_1}{x_2 - x_1}$$

This formula can be summarized as "rise over run," where "rise" is the change in the y-values and "run" is the change in the x-values.

The "rise over run" method suffers because it really only uses two points on the line. A better method is to include *all* the data points in a linear regression. This is a statistical procedure that minimizes the sum of the squares of the distance from the actual data to the line (which is why it is often called the "least squares" method). Regression analysis is available in many hand-held calculators and computer graphing methods. Refer to your manual for directions.

If the data points do not lie on or close to a line but do smoothly change as the values of the variables change, then the value of the dependent variable has a *non-linear dependence* on the independent variable. It is difficult to determine by eye what mathematical function applies in these cases. Although there are computer programs that can "fit" a curve with different functions, the most meaningful procedure is to try several different functions by regraphing the data with the y-values transformed. For example, in some experiments there is a logarithmic dependence between the dependent and independent variables. In that case, we would find that if we graph ln y on the y-axis we get a line.

Calculating Unknowns from Graphs

Many experiments use graphs of known data points to get information about an unknown quantity. This is most important with calibration curves, which are created from data for known values of an independent variable and an observed value of a dependent variable. A good calibration curve can be used to determine what the "unknown" independent variable is in a sample. This is done by obtaining a reading for the dependent variable and then locating the value of the dependent variable that corresponds to that reading.

In using calibration curves, it is critical that you *interpolate* the data if at all possible. This means that the curve should only be used on values that are *within* the "known" data points. If, for example, we found in the experiment listed in Figure I-4 that an unknown sample had and absorbance of 0.50, we would be able to determine that the concentration of dye in that sample had a value of 4.0×10^{-5} mol/L.

D. Analysis of Chemical Reaction Rates and Reaction Kinetics

In the laboratory, you will be studying reactions in a "batch" mode, with a fixed amount of reactants in a solution. Real engineering settings, however, usually work in a continuous flow mode, in which a solution of a substance is moving through a reactor that contains other ingredients for the reaction. Although continuous flow experiments must always be done before a reaction is fully understood, batch mode studies can give important information about the kinetics of a reaction, also.

The kinetics of a decomposition reaction are among the simplest to study, because there is only one reactant:

$$a\text{A} \rightarrow b\text{B} + c\text{C} + \cdots$$

The rate of these reactions can be studied by following the production of the products or the consumption of the reactant. Rate is generally expressed as a change in concentration (usually molarity, mol L^{-1}) per unit time (usually seconds). To make valid comparisons among the different concentration changes, it is necessary to divide each concentration change by the stoichiometric amount of product:

$$\text{rate} = \frac{1}{b} \frac{\Delta[\text{B}]}{dt} = \frac{1}{c} \frac{\Delta[\text{C}]}{dt}$$

We can also express the rate as a function of the decrease in the concentration of a reactant. Since the concentration of a reactant decreases with time, changes in reactant concentration are negative. Therefore, we include a negative sign in the equation.

$$\text{rate} = -\frac{1}{a}\frac{\Delta[A]}{dt}$$

Rate Laws

The rate of a chemical reaction depends on many factors, including the concentrations of chemical substances in the reaction mixture. This is generally expressed as a rate law, which relates the rate to the concentrations of substances:

$$\text{rate} = k[A]^x[W]^y$$

The rate law is experimental. We have to observe how the rate depends on concentration.

Here, A is the reactant in the decomposition reaction and W is the concentration of some other species. W can be another reactant. W can even be a substance that is not involved in the stoichiometry of the reaction. In that case, W is referred to as a catalyst or as an inhibitor.

As an example of the derivation of a rate law, consider the data shown in this table:

Time (s)	10 s	30 s	60 s	120 s
Slope (mol L^{-1} s^{-1})	−0.0050	−0.0045	−0.0035	−0.0022
Rate (mol L^{-1} s^{-1})	0.0050	0.0045	0.0035	0.0022
[A] (mol L^{-1})	0.648	0.555	0.441	0.277

The rate decreases as [A] decreases. When [A] drops from 0.648 M to 0.441 M (a change of −0.207, or −32%), then the rate decreases from 0.0050 to 0.0035 (a change of −0.0015, or −30%). The relative decrease in the concentration of A and the rate of the reaction are nearly equal, and we conclude that the rate law has an $[A]^1$, or simply an [A], term.

Thus, we can write, for just [A]: rate = $k'[A]$. This gives us the chance to determine the rate constant at some concentration of [A], k' = rate/[A]. Choosing the data point for 30 s, when [A] = 0.555, we get k' = 0.0045 mol L^{-1} s^{-1}/0.555 mol L^{-1} = 0.0081 s^{-1}.

This works fine for [A] and other substances in the reaction. But what about reactions where the rate depends on substances that are *not* involved in the reaction? Such substances, if they speed up a reaction, are called *catalysts*. Catalysts are not consumed or made in a reaction. Therefore, their concentration stays the same during the reaction.

We evaluate the order of a reaction for a catalyst by examining the rate as we change the amount of the catalyst. Let's say we derived the dependence of rate on [A] with 0.0025 M catalyst W present. But what if we increase the amount of W to 0.0050 M? We then might get data that look like those in this second table:

Time (s)	10 s	30 s	60 s	120 s
Slope (mol L^{-1} s^{-1})	−0.0093	−0.0074	−0.0046	−0.0022
Rate (mol L^{-1} s^{-1})	0.0093	0.0074	0.0046	0.0018
[A] (mol L^{-1})	0.600	0.441	0.277	0.110

By comparing the data, we can make two important observations. First, the reaction with 0.0050 M W is much faster. Note that it now takes only 30 s to reach a [A] = 0.441 M, whereas in the first case it took 60 s: doubling the amount of W doubles the rate. Second, note that the rate when [A] = 0.441 M increases to 0.0074 mol L^{-1} s^{-1} from 0.0035 mol L^{-1} s^{-1}. The rate of the reaction is twice that in the first case. This shows the reaction rate is dependent on [W]^1, or simply [W].

We now can write the rate law:

$$\text{rate} = k[A][W]$$

Information about the dependence on [W] can also be obtained by getting the rate constant k' for this second set of data. At 30 s, [A] = 0.441 mol L^{-1} and k' = 0.0074/0.441 = 0.0167, twice the value at [W] = 0.0025. This also indicates that the rate is linearly dependent on [W].

Integrated Rate Laws

Mathematical calculus, which we will treat as outside the scope of this course, provides a final, and most powerful, method for determining the rate law. Let us return to the case of a rate law that depends on a single concentration, or on a single concentration that varies during a reaction. In that case, we can use calculus to determine the concentration of the species A as a function of time.

Reaction Order	Rate Law	[A] vs. t Dependence	Characteristic Plot
Zero	rate $= k'$	$[A] = [A]_0 - kt$	[A] vs. t Slope is $-k$
First	rate $= k'[A]$	$[A] = [A]_0 e^{-kt}$ $\ln([A]) = \ln([A]_0) - kt$	$\ln([A])$ vs. t Slope is $-k$
Second	rate $= k'[A]^2$	$1/[A] = (1/[A]_0) + 2kt$	$1/[A]$ vs. t Slope is $2k$

Figure I-5 shows three plots for the data tabulated earlier.

Only one of the plots shows a straight line: the middle one. This indicates that the reaction is also first order in [A]. And a comparison of the two slopes confirms that the reaction is also first order in [W].

FIGURE I-5

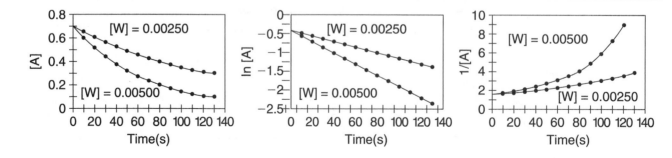

Designing a Kinetic Study

With these principles in mind, we can now outline four steps in designing a kinetic study:

1. Carry out the reaction in a manner that permits the periodic determination of a reactant or product concentration. Make sure that the experiment allows for fast and complete mixing of reactant solutions.

2. Collect the data until at least 25% or more of the reaction has proceeded. Repeat the data collection until it is reproducible.

3. Convert the experimental data into concentrations of a reactant. Calculate $[A]$, $\ln[A]$, and $1/[A]$ and then graph each vs. time to see which fits a line best. Determine k for the reaction from the best fit line.

4. If a catalyst or other "unreactive" substance is present, vary it and see how k for the reaction changes with time.

Homogeneous and Heterogeneous Catalysis

A catalyst increases the rate of reaction without being consumed. If the catalyst is a dissolved salt such as a solution of potassium iodide, the reaction is catalyzed homogeneously. The rate can be expressed as

$$\text{rate} = k[W][A]^x$$

and $[W]$ can be given in concentration units.

Other catalysts are insoluble solids, like MnO_2. In this situation, reactant molecules from the liquid phase adsorb onto the surface of the solid particles. There they can decompose and the parts can scramble around and reattach to other fragments. The products then desorb from the surface. The overall process can be termed a catalytic sequence.

When the catalyst maintains a separate phase from the reactants and products, it is termed heterogeneous catalysis. This class of reactions, especially solid catalyzed gas phase reactions, is in fact the most prevalent in industry. A common example is the detoxification of engine exhaust gases which occurs over the solid catalyst in a car's catalytic converter.

When a solid catalyst is employed it is common to write the rate with units of moles reacted per gram of catalyst per time,

$$\text{rate (mole g}^{-1}\text{ s}^{-1}) = k''\,[W][A]^x$$

where $[W]$ is now in grams per liter and for $x = 1$, k'' has units ($L^2\,g^{-2}\,s^{-1}$). Or, since $[W]$ is constant it can be combined with k'' to give

$$\text{rate (mole g}^{-1}\text{ s}^{-1}) = k'''[A]^x$$

where the units of k''' are ($L\,g^{-1}\,s^{-1}$). When performing economic calculations involving catalyst cost, you'll want to keep the $[W]$ term separate from k''.

A few additional notes can be made about catalyzed reactions. First, it is not uncommon to notice lag times at the beginning of a batch experiment employing a fresh catalyst. It might take some time for the catalytic material to attain its "active" form—for example, through a change in the oxidation state of a dissolved catalyst, or the removal of unreactive material from the surface of a solid catalyst. The activation phenomenon must be accounted for in the rate measurements for an accurate measurement of catalytic activity.

Second, increases in the rate will occur over a solid catalyst if the catalyst particles break up during the reaction. If fracture occurs, more catalytic surface is exposed, even though the mass of catalyst remains the same. Many expensive catalyst materials are made in the form of small particles for this reason. It is important to keep the amount of active area constant during an experiment if catalyst activity is being measured.

ACKNOWLEDGMENTS

The scenario for Experiment Group B and guidance for the adaptation of clinical methods to measure and treat unhealthy serum iron levels were provided by Prof. John Fitzloff of the College of Pharmacy at the University of Illinois at Chicago. Editorial assistance by Prof. Barbara Weil from William Rainey Harper College is gratefully acknowledged.

The scenario for Experiment Group C and guidance for the adaptation of the BCG method for determination of serum albumin were provided by Prof. Chuck Woodbury of the College of Pharmacy at the University of Illinois at Chicago. Editorial assistance by Dr. Audrey Hammerich is gratefully acknowledged.

The scenario for Experiment Group D and guidance for the adaptation of leaf compost chambers for CO_2 analysis from the decay of leaves were provided by Prof. John Lussenhop of the Department of Biological Sciences at the University of Illinois at Chicago. Editorial assistance by Dr. Helene Gabelnick from Harold Washington College is gratefully acknowledged.

The scenario for Experiment Group E and guidance for the description of pH methods in medicine, were provided by Prof. Janean Holden of the College of Nursing at the University of Illinois at Chicago. Comments about the procedure by Dr. Dennis Lehman from Harold Washington College are gratefully acknowledged.

The scenario for Experiment Group F was drafted by Prof. John Regalbuto in the Department of Chemical Engineering at University of Illinois at Chicago. Editing support by Prof. Luke Hanley of University of Illinois at Chicago is gratefully acknowledged.

The scenario for Experiment Group G originated in discussions with Joan Radtke MS MT(ASCP)SC, CLS(NCA) of the School of Biomedical and Health Information Sciences in the College of Health and Human Department Sciences at the University of Illinois at Chicago.

We gratefully acknowledge the assistance of Dr. David M. France in the development of the scenario of Experiment Group H and for providing background information on the factors engineers might consider in the design of a fire-resistant box. Dr. France is Associate Dean of Engineering and Professor of Mechanical Engineering at the University of Illinois at Chicago.

The scenario for Experiment Group I was drafted by Prof. John Regalbuto in the Department of Chemical Engineering at the University of Illinois at Chicago.

The scenario for Experiment Group J and guidance for the adaptation of procedure to analyze for the mineral nitrogen content in soil were provided by Prof. John Lussenhop of the Department of Biological Sciences at University of Illinois at Chicago.

EXPERIMENT GROUP A

METHODS OF INQUIRY AND MEASUREMENT: FLOAT AN EGG

The density of a solution affects whether an object floats in that solution.
Jeff J. Daly/Visuals Unlimited.

PURPOSE These experiments will introduce you to the basics of mass and volume measurement in a chemical laboratory. They are all built upon some student inquiry, so they also introduce inquiry methods. Finally, they will provide you the opportunity to work individually and as a member of a group.

SCHEDULE OF THE LABS **EXPERIMENT 1: "Physical Properties: The Glass Bead Lab"**

1. Determine the volume, mass, and density of an individual glass bead (individual work).
2. Combine beads with a group of students and determine their average volume, mass, and density (group work).

EXPERIMENT 2: "Use of Volumetric Glassware: The Floating Egg Problem"

1. Prepare a solution to float an egg (group work).
2. Determine solution density using volumetric glassware (individual work).

EXPERIMENT 3: "Chemical Proportionality: Carbonate and Hydrochloric Acid"

1. Characterize the reaction of a carbonate with acids.
2. Determine the mass-to-volume ratio for the reaction of a carbonate with hydrochloric acid.

Experiment 1
PHYSICAL PROPERTIES: THE GLASS BEAD LAB

Pre-Laboratory Assignment   **Due Before Lab Begins**

NAME: _____

Complete these exercises after reading the experiment but before coming to the laboratory to do it.

1. Identify the following properties of calcium metal (listed in the reference book *The Merck Index*) as either chemical or physical properties and as intensive or extensive properties.

   a. Lustrous silver-white surface (when freshly cut).

   b. Ignites in air when finely divided, then burns with crimson flame.

   c. Acquires bluish-gray tarnish on exposure to moist air.

   d. Boiling point is 1440°C.

   e. Melting point is 850°C.

   f. Reacts with water, alcohols, and dilute acids with evolution of hydrogen.

2. Calculate the volume of a solid sphere if its circumference is 5.5 cm.

3. Determine the density of the solid sphere from question 2 if it has a mass of 7.421 g.

4. You are not allowed to eat or drink in the lab at any time, including during this experiment even though you are working with only glass beads and water. Explain why you are not allowed to eat and drink in the lab.

Experiment 1
PHYSICAL PROPERTIES: THE GLASS BEAD LAB

BACKGROUND

Chemical and physical properties are used to characterize and to help identify a substance. A chemical property describes how a substance interacts with other substances and involves a change in the chemical composition when the property is observed. An example of a chemical property of sodium metal is its reaction with water to produce sodium hydroxide and hydrogen gas. A physical property describes the substance itself, and no change in chemical composition occurs when the property is observed. Examples of physical properties of sodium metal include melting point, 97.82°C; boiling point, 881.4°C; density, 0.968 g cm^{-3}.

Properties can be further categorized as extensive or intensive properties. Extensive properties depend upon the amount of matter present; intensive properties are independent of the amount of matter present. For example, the length of a piece of copper metal depends on the amount of matter present and so is an extensive property. However, the temperature at which water boils is 100°C at standard atmospheric pressure regardless of the amount of water present; boiling point therefore is an intensive property.

PURPOSE

You will work individually and as part of a group during this experiment. Your goals are to measure the mass and volume of a single glass bead, and then to calculate its density. You will also determine the average mass and volume of a set of glass beads, and calculate the average density for the set of beads. You will practice some basic measurement techniques and error analysis, and you will characterize some properties of a substance. In addition, you will be introduced to inquiry methods in chemistry: You will decide which procedures to use in certain cases.

PROCEDURE

Part I: Formation of Groups

This experiment is best done in groups of three or four individuals. Determine, with your group, how to obtain the needed measurements in order to calculate the volume of a single glass bead. Once you have agreed upon a procedure to determine the volume of a bead, work individually as outlined in Part II. Then reconvene as a group to complete the experiment in Part III.

Here is some potentially useful information: The radius, r, of a sphere is the distance from its center to any point on its surface; the diameter, d, is twice the radius; the circumference, C, of a sphere is the distance around the sphere; and the volume, V, is the total amount of space the sphere occupies. Mathematically,

$$d = 2r$$

$$C = 2\pi r$$

$$V_{\text{sphere}} = 4/3(\pi r^3)$$

Part II: Determination of the Density of a Single Glass Bead (Individual Work)

The density of a substance is the proportion that relates mass and volume:

$$d = m/V$$

To calculate the density of the glass bead, you need to measure the mass and volume of the bead. Obtain a single glass bead. Measure its mass on a balance. Determine the volume of the glass bead using the procedure decided upon by your group. Be sure to record this procedure in your lab notebook.

Using your values for mass (m) and volume (V), calculate the density (d) of the glass bead in grams per cubic centimeter.

Repeat your measurements two more times. Tabulate all of your data. Determine an average mass, volume, and density, including average deviations for your individual bead.

Part III: Determination of the Density of a Set of Glass Beads (Group Work)

Reconvene your group and add each of the glass beads to a set of additional glass beads from your instructor.

Obtain the mass of the entire set of glass beads.

To determine the volume of the set of glass beads, partially fill a graduated cylinder with water and record this initial volume of water in your lab notebook. Add the glass beads. All of the glass beads must be completely submerged in the water. Record the final volume of water and glass beads. The difference between these two measurements is the volume of the glass beads. This method of volume determination is the displacement method.

Determine the density of the glass beads. Repeat your measurements two more times. Report an average mass, volume, and density for your set of glass beads. Include average deviations.

REPORT

Describe the procedure you used to determine the volume of the individual glass bead. Tabulate the masses, volumes, and densities of the individual glass beads for each person in your group. Then tabulate the masses, volumes, and densities of the set of glass beads. Explain which average density you think is a more reliable measure of the density of the glass beads, that of the individual bead or that of the set of beads. Compare the masses, volumes, and densities of the individual glass beads with each other and with the set of glass beads. Identify each of the three properties (mass, volume, and density) as either intensive or extensive. Support your answers with your experimental data.

QUESTIONS TO ANSWER IN YOUR REPORT

1. Distinguish between the mass and the weight of a material.

2. Why is the displacement method used to determine the volume of the set of beads instead of simply pouring the set of beads directly into the empty graduated cylinder?

Experiment 2
USE OF VOLUMETRIC GLASSWARE: THE FLOATING EGG PROBLEM*

Chemists use special **volumetric** glassware in the chemistry laboratory (rather than the ordinary beakers and flasks) when they need high-precision measurements of the volumes of liquids. Understanding the proper use and limitations of such glassware is important in achieving reliable results. Experimental technique and the inherent accuracy of the glassware both affect experimental results. One of your personal goals for this course is to develop sound experimental techniques.

In this experiment you will practice with volumetric glassware to improve your technique and to solve an important historical problem in practical chemistry. Volumetric glassware is designed for two quite different purposes:

- To deliver (TD) an accurate volume of liquid.
- To contain (TC) an accurate volume of liquid.

A container meant to *contain* an accurately known volume could of course be used to deliver the liquid to another vessel. However, the amount actually delivered would never exactly equal the amount originally in the container. Some liquid would inevitably remain behind, clinging to the interior walls. Some liquid also remains behind on the walls of a piece of glassware that is TD designed, but the calibration of the TD glassware takes this into account.

An Historical Question

For many years soap was made at home from a variety of recipes. Animal fat, usually from cattle (also called tallow) was cooked with a lye solution. Lye, though it was mostly simple sodium hydroxide, could not be made from purified chemicals, as we do now. Instead, the solution was obtained from ashes and water. The ashes were treated with hot water, and then the mixture was filtered to obtain a solution. But before this could be used in soap making, one had to check the concentration of the lye solution. One simple test was to try to just float a raw egg in the solution. If the egg sank, the concentration of the lye in the solution was too low. If the egg floated too high, the concentration was too great, and water was added before adding the fat. To "just float" in this case means to make the top of the egg just touch the top of the solution, without any significant amount of the egg protruding above the surface of the solution.

PURPOSE

You are to determine the density (in grams per milliliter) of the solution needed for the production of soap. In addition, you are to determine in what way each of the following affects the value and the accuracy of the calculated density: (a) the type of volumetric glassware used (volumetric flask, Mohr pipet, volumetric pipet, or buret), and (b) the degree of freshness of the raw egg.

---

* Portions of this laboratory, including the "historical question" are adapted from materials developed by Dr. Susan Hershberger in the Miami University Middletown in their Scenario Lab program.

PROCEDURE

Part I: Formation of Groups and Preparation of the Solution

This experiment is best done in groups of three or four individuals. Follow the detailed directions of your laboratory instructor.

Because lye is caustic and corrosive, the substitute sodium chloride will be used in this lab. The solutions are still concentrated enough to pose a hazard to the eyes. Goggles must be worn. Salt, eggs, standard glassware, balances, and deionized water will be available. Other items may be available on request.

Clean and dry all glassware before use. See the Introduction for more information.

Follow these guidelines to prepare a solution that will "just" float an egg:

Make at least 1500 mL of the solution

Use only one egg for testing. Note the code. At the end of the experiment, you will use this to determine the freshness of the egg.

When you have prepared the solution, each individual should record the procedure in his or her notebook.

Part II: Measurement of Solution Density

You will do this part individually. Each individual should use all four kinds of measuring devices.

The group should obtain a 25-mL buret, a 25-mL Mohr pipet, a 25-mL volumetric pipet, and a 25-mL volumetric flask. (*These must be cleaned and returned to their designated location at the end of the experiment.*)

Decide who will work first with each type of equipment. For the equipment that dispenses solution (pipets and burets), you will determine the mass of the solution by collecting the solutions into a weighed Erlenmeyer flask. For the volumetric flask you will determine the mass of the solution in the flask itself. So, obtain the weight of a clean, dry Erlenmeyer flask and of the volumetric flask. Don't forget to take a notebook to the balance, too. Before weighing, be sure the balance reads zero when nothing is on the pan. Check with your instructor if this is not the case. Then place the flask on the balance. This is the mass of the empty container.

Record the sensitivity of the balance in your notebook. Note for future reference that when the net mass of something is determined by subtracting the empty (tare) mass from the gross mass, *the uncertainty will be twice the sensitivity.* This is because there is no guarantee that each of the two mass readings will deviate from the true mass in the same manner (i.e., both larger or both smaller).

Now, carry out three repetitions of a determination of the density of the solution with each piece of glassware by measuring a volume close to 25 mL of the solution and then determining the weight of the flask plus the solution.

A. Buret

Make sure the stopcock on the buret is properly assembled. Clean the buret, and make sure that it drains smoothly with no beading of water on the walls. Then pre-rinse it with a little of your solution. Use a buret clamp and ring stand to mount the buret vertically. Fill it to about the zero level with your solution. Drain some out to ensure there are no pure water or air gaps in the tip of the buret. Record the level reading after it has stabilized.* The level should be recorded to the nearest 0.01 mL,

---

* Readings that are obtained too quickly do not include the liquid that slowly trails down the sides of the buret.

but it does not have to be at exactly 0.00 mL. (Estimate the location of the bottom of the meniscus between the calibration marks, which are 0.1 mL apart.)

Deliver solution from the buret into the clean, dry, weighed Erlenmeyer flask. Deliver until the buret level is near the 24-mL mark. Record the reading after waiting for the level to stabilize.

Now weigh the flask plus the solution and record the total mass. When you subtract the mass of the empty, dry flask, you will obtain the mass of the solution.

Repeat the entire procedure, including rinsing the buret, for a total of three trials. You want to develop your technique so that you can use a buret accurately and reproducibly (precisely).

B. Mohr Pipet

Never apply suction by mouth to fill up this or *any* pipet. Pipet bulbs will be available in the laboratory.

Using the 25-mL Mohr pipet, carry out the weighing procedure of Part A. Be sure that the pipet is clean and drains correctly, then pre-rinse it with some of your solution. Attempt to deliver exactly 23 mL of solution into the dry weighed Erlenmeyer flask. If you undershoot or overshoot, record whatever actual volume you end up delivering. Do a total of three trials. Attempt to perfect your technique.

C. Volumetric Pipet

Using the volumetric pipet, carry out the weighing procedure of Part A using a clean, pre-rinsed 25-mL volumetric pipet to deliver 25.00 mL of solution. If the pipet is marked "TD," it is designed to give 25.00 mL by draining with gravity: Do not include the solution that remains in the tip after draining. The tips of the pipets are very fragile. Be careful not to exert pressure on them. Do a total of three trials.

D. Volumetric Flask

Fill your pre-weighed volumetric flask with the solution, stopping at the calibration mark. Weigh the solution-filled container and record the mass. This is the mass of the flask plus the solution "contained," and, by subtracting the mass of the empty flask, you can determine the mass of the solution contained. *Empty* the contents of the container into the clean, dry, weighed Erlenmeyer flask. Record the mass of the Erlenmeyer flask plus the solution. From this you can calculate the mass of solution "delivered." Repeat this procedure three times.

Cleanup.   The glassware should all be rinsed well with deionized water before returning it to its designated location; unrinsed glassware will dry out to leave solid sodium chloride that may clog the tips. Also, return the eggs to the proper container for that code. Do not put all the eggs in one basket.

REPORT

Your report should include a full procedure and analysis of your data. This analysis should include an indication of the variation among the measurements. Also, present a comparison of your data to those obtained by the others in your group and by others in the room with the same glassware. How did the amount of variation in your data compare to the amount obtained by others with the same or with different glassware? Finally, answer these questions in one or two sentences:

1. What was the density of the solution in which the egg just floated?
2. How valid is the assumption that aqueous sodium chloride is an adequate replacement for aqueous sodium hydroxide in this problem? Explain your reasoning.

3. Compare the density you obtained with the densities obtained for different eggs. What, specifically, was the effect of the freshness of the egg?

4. What were the advantages/disadvantages of each measuring device?

5. Compare the mass of the solution "contained" in the volumetric flask to the mass of the solution "delivered" by the volumetric flask. Account for differences. Which of the two masses, contained or delivered, should you have used in the determination of the density of the solution? Why?

Experiment 3
CHEMICAL PROPORTIONALITY: CARBONATE AND HCl*

BACKGROUND

Some of the basic principles of chemistry were recognized and used well before the formulation of atomic theory in the early nineteenth century. Many of these practical procedures are still employed today. For example, some food preparation requires careful control of the amount of baking soda (sodium hydrogen carbonate) or baking powder (a mixture of tartaric acid and baking soda). Other examples involve the mixture of ingredients needed to make a good cement, or the proper application of fertilizers to adjust soil conditions. In all of these, there is an implicit rule: We can determine how much to add by considering the relative proportions of the components. This lab shows how an early method for determining acid concentration used such proportional reasoning.

An Historical Question

In the eighteenth century, tanners developed a simple method to determine the concentration of the acid they used in the tanning process. They added small portions of a solid carbonate salt to the acid solution, this resulted in fizzing as the carbon dioxide gas generated by the reaction escaped from solution. They continued adding the carbonate salt until there was no more evidence of fizzing. If they used a set volume of acid, they could compare the amount of salt required for various samples of acid and thus "measure" them one against the other. Acid samples that required large masses of carbonate salt were more concentrated than acid samples that required small masses of carbonate salt.

PURPOSE

In this lab, you will investigate what occurs when a solid substance, a metal carbonate, is added slowly to a solution of an acid. You should be able to validate the proportionality of the amounts—volume and mass—that are required to accomplish this. This proportionality underlies the idea of chemical stoichiometry.

By completing this experiment you will

- Gain experience making observations.
- Gain experience in designing procedures for a chemistry laboratory.
- Recognize some easily observable changes that signal a chemical reaction.
- Investigate proportional reasoning applied to mass and volume measurements.

Solutions are defined as homogeneous mixtures of solutes and solvents. In this course we will be using many solutions, most of them water-based or aqueous solutions in which water is the solvent. The amount of solute dissolved in a specified volume of water is given by the concentration: Dilute solutions contain small amounts of dissolved solute, and more concentrated solutions contain greater amounts of dissolved solute.

---

* Portions of this laboratory, including the "historical question," are adapted from materials developed by Purdue University.

How do we measure the concentration of a solution? We can use many different methods, some of which are more accurate than others. We can use qualitative information that is based primarily on the physical properties of the solution. Colored solutions can be classified as more or less concentrated based on their depth of color. Hygrometers can measure the specific gravity of a solution and thereby distinguish between dilute and more concentrated solutions. These are all physical measurements.

We can also use chemical methods to determine concentration. These methods involve a chemical reaction between some reagent and the dissolved solute. Three things are required here: (1) We must carefully measure the amount of reagent added, (2) we must have some means to know when to stop adding the reagent, and (3) we must be able to relate the amount of reagent added to the chemical reaction being monitored. In addition, we must rely on the proportionality of the system—the relative amounts of mass and volume—being the same regardless of the initial mass or volume.

In the experiment you will do today, a metal carbonate is added to the solution until there is no more evidence of carbon dioxide being generated. The chemical equation for this reaction is

$$M_2CO_3(s) + 2\ HCl\ (aq) \rightarrow 2\ MCl\ (aq) + CO_2\ (g) + H_2O\ (l)$$

You can monitor the evolution of CO_2 by observing what happens as CO_2 escapes from solution. To monitor the amount of metal carbonate required by the reaction, you can simply subtract the mass of carbonate you have at the end from the original mass of carbonate you had at the beginning. The relationship between the chemical amount (usually, the number of moles) of metal carbonate and that of the HCl is given by the stoichiometry of the equation, but, because we are working with mass measurements and unknown solution concentrations, we must *experimentally* determine the mass-to-volume ratio.

This will require some careful note taking. The stated goal is to determine the mass of metal carbonate required to completely react with a particular volume of solution. So, you have two kinds of measurements to record: mass and volume. Mass is easily obtained by using a balance. For this experiment, you will use the analytical balance, which will allow more precise measurements than the top-loading balances. Review the proper techniques for using the analytical balance. Do not bring any opened containers of chemicals into the analytical balance room. Instead, weigh a clean, dry beaker, then return to the laboratory. Use the centigram balances to measure approximately 1 g of metal carbonate into the beaker, then return to the balance room and weigh again. The difference between the two masses is the mass of your initial supply of metal carbonate.

At the end of the experiment you will again weigh this beaker with the remaining metal carbonate. This is called "weighing by difference" and will allow you to determine the mass of metal carbonate actually consumed in the experiment. Report all masses to 0.001 g. Use graduated cylinders to measure solution volumes. The most accurate volumes are read by holding the cylinder at eye level and estimating the volume marked by the bottom of the meniscus.

PROCEDURE

Part I: Formation of Groups

This experiment is best done in groups of three or four individuals. Follow the detailed directions of your laboratory instructor. As a group, obtain a small amount of sodium carbonate to be used in Part II. The goal of Part II is to distinguish whether or not a gas is evolved when metal carbonate is added to water and to various water solutions. After completing Part II, develop, as a group, a work plan to complete the remainder of the experiment.

Part II: Dissolution or Chemical Reaction?
What Do Your Observations Mean? (Group Work)

1. Using a spatula, add a small amount of sodium carbonate to about 10 mL of water. Swirl the solution so that the solid has an optimum chance to dissolve. What observations can you make? Does the solid in fact dissolve? Do you sense any other changes? No other physical change (yes)

2. Add a small amount of sodium carbonate to about 10 mL of the hydrochloric acid solution and swirl gently. Does the solid dissolve? Do you sense any other changes? (yes) Bubbled

3. Add one spatula of sodium carbonate to about 10 mL of dilute acetic acid solution and swirl gently. What observations can you make about this combination? (Notice that the directives here are less specific than in Parts 1 and 2. This was intentional. The detailed directions teach you experimental theory and procedure; the mere outlines of the system to be studied allow you to apply the procedures learned.) Dissolved no change

Part III: Measuring the Strength of a Solution (Individual Work)

This part of the procedure should be divided up among the individual group members according to the work plan decided upon by the group.

The group must determine the mass of sodium carbonate required to completely react with a measured volume of acid. You have two acid solutions of different concentrations to examine. The total volume of each solution that is available is limited to about 100 mL. Each group will perform four trials with different volumes for each acid solution, for a total of eight trials.

Part IV: Collection and Analysis of Group Data (Group Work)

The next step in reporting results for any experiment is to critically examine both the procedure and the experimental results. Before you leave, reconvene with the whole group to be sure you have each collected the necessary data. Carry out a preliminary analysis together. Failure to do this may mean that you do not have the required information to prepare your report.

Prepare tables of data for both solutions A and B. Each table should contain the four solution volumes your group used and the mass of sodium carbonate required for each volume. You will use these tables to analyze your data.

Do the following for solution A and for solution B, separately.

1. Compute the ratio of mass to volume for each of the four trials. Use units of grams and milliliters. Determine the average value of this ratio.

2. Prepare a graph of mass versus volume. Determine the slope of the best line drawn through the data points. (If you use a computer to prepare this graph, you may use a linear regression analysis to get the slope; if you do, report the correlation coefficient the program reports).

Part V: Analysis of the Molar Mass of an Unknown
Metal Carbonate (Individual Work)

As a group you have prepared two graphs of the reaction of sodium carbonate, Na_2CO_3, with solutions of hydrochloric acid. You can now use these to determine whether another substance is lithium carbonate, Li_2CO_3, calcium carbonate, $CaCO_3$, or potassium carbonate, K_2CO_3.

Ask your instructor for directions in obtaining an unknown. This will be a coded sample in a small vial. You will only get one portion of this, so use it carefully.

Obtain the mass of the vial with your unknown to the nearest 0.001 g.

In the lab, put 25 mL of one of the hydrochloric acid solutions into a clean beaker. Note which acid solution you use. Begin adding the metal carbonate to the hydrochloric acid. Do this very carefully, to determine exactly how much carbonate is needed for the reaction. Once you are certain the hydrochloric acid is completely reacted, put the cap back on the vial of the unknown and reweigh it.

You can use your data and the data collected by the group to determine the molar mass of your unknown. Your group data will tell you how many *grams* of *sodium carbonate* react with 25 mL of the solution of HCl you used. You can do this even if your group did *not* do an experiment with 25 mL of HCl solution: You can read the graph to determine the mass of sodium carbonate needed for that volume of HCl.

This mass of sodium carbonate is equivalent to a mole amount of sodium carbonate. We can convert using the molar mass of sodium carbonate:

$$x \text{ g Na}_2\text{CO}_3 \times \frac{1 \text{ mol Na}_2\text{CO}_3}{105.988 \text{ g Na}_2\text{CO}_3} = y \text{ mol Na}_2\text{CO}_3$$

This is the same number of moles of *your* carbonate that reacted with 25 mL of the hydrochloric acid. You can use this mole amount and the mass of your carbonate needed for this to get the molar mass of your carbonate:

$$\text{moles of unknown carbonate} = \text{moles of Na}_2\text{CO}_3$$

$$\text{molar mass of unknown carbonate} = \frac{\text{mass of unknown carbonate}}{\text{moles of unknown carbonate}}$$

Report the molar mass of your unknown carbonate to the nearest 1 g mol^{-1}.

Finally, you can determine the composition of the unknown carbonate by comparing the molar mass you calculate with the molar masses of the three possible carbonates, Li_2CO_3, CaCO_3, and K_2CO_3.

REPORT

Summarize the procedure and your group's work plan for this experiment. Comment on any difficulties with the procedure as a means to accomplish the goal of the experiment.

Your report must also include the data tables for solutions A and B, the average mass-to-volume ratios, the slopes of the lines from the graphs of mass versus volume (and the correlation coefficient if you used a statistical program to analyze the data), and the identification of the unknown carbonate. Include sample calculations for any calculated values.

QUESTIONS TO ANSWER IN YOUR REPORT

1. For each acid solution, how closely do the individually computed mass-to-volume ratios agree with one another?
2. For each acid solution, compare the slope of the line with the average value obtained for the computed mass-to-volume ratios for that solution.
3. Compare the average of the computed mass-to-volume ratios of solution A to that of solution B. Which is more concentrated? What is the relative ratio of the concentrations of the solutions?

4. Compare the slope of the graph obtained for solution A with the slope obtained for solution B. Which is more concentrated? What is the relative ratio of the concentrations of the solutions?

5. Which ratio do you think is more accurate for each solution, the average value of the computed ratio or the ratio obtained from the slope of the line? Why?

EXPERIMENT GROUP B

METAL IONS AND THE BLOOD: TREAT IRON DEFICIENCY . . . AND OVERLOAD

Spectrophotometers are an essential part of many clinical laboratories.
Courtesy of Foss NIR Systems, Inc.

PURPOSE   This experiment group introduces you to the way that we can determine the contents of a solution by measuring how it interacts with light. This kind of analysis is called spectrophotometry—the measurement (metering) of the light absorbed (the spectrum) by a solution. As the weeks progress, you will have more and more responsibility for designing the actual procedures.

In the case of this set of experiments, we have borrowed material from the department of medicinal chemistry in a college of pharmacy: They use spectrophotometry to determine the amount of iron in blood, part of a diagnosis and treatment. We will not have you work with real blood, but in the third experiment you will carry out procedures that emulate the actual methods.

SCHEDULE OF THE LABS

EXPERIMENT 1: Skill Building Lab: "Spectrophotometry of Dyes"

1. Prepare a standard solution of a food dye (individual work).
2. Determine its UV–visible spectrum (individual work).
3. Combine your solution with other solutions to determine the changes in color and spectrum that result (group work).

EXPERIMENT 2: Foundation Lab: "Spectrophotometric Determination of Cu^{2+}"

1. Construct a calibration curve using standard solutions (group work).
2. Determine the concentration of Cu^{2+} in an unknown solution using the calibration curve (individual work).

EXPERIMENT 3: Application Lab: "Total Serum Iron Assay"

1. Construct a calibration curve using standard solutions to match the iron concentration range that might be found in adult males, adult females, and children (group work).

2. Measure the total serum iron concentration in a "patient's blood sample" (individual work).

3. Treat the patient and measure the serum iron level after treatment (individual work).

SCENARIO

You work in a clinical laboratory. A common assay used to aid in the diagnosis of diseases affecting the blood (or serum) is for total serum iron. An assay is a test for an amount of a substance. Because of the importance of iron in many biological processes, several health problems arise from levels of iron that are too low or too high.

Iron in food supplements is usually consumed orally in the +2 oxidation state, but in the body it oxidizes to the +3 oxidation state. The Fe^{3+} ions attach to ferritin, a large iron-storage protein (see Figure B-1). Fe^{3+} is released from the ferritin to transferrin, another protein present in the blood plasma, which transports the iron ions to blood forming sites.

Two blood samples have just arrived at the lab. One is from a woman who has been experiencing fatigue, weakness, shortness of breath, low blood pressure, and headaches. The doctor suspects iron deficiency anemia. Another is from a child who has taken several of his mother's iron supplement pills, thinking they were candy. You receive one of the two samples. Your job is to measure the serum iron level in your sample and report the results to the doctor and pharmacist. They would then prescribe and dispense the necessary treatment to raise the iron level of the woman and lower that of the child. You will temporarily assume the role of the doctor here and "treat" the patient by either increasing or decreasing the iron concentration in your sample.

After some time (minutes in this emulation, days in practice) more tests would determine whether the patient responded to treatment. You will then resume the role of medical technician and measure the serum iron in the treated patient's sample.

FIGURE B-1 STRUCTURE OF THE PROTEIN TRANSFERRIN. IRON BINDS AT SEVERAL PLACES IN BOTH HALVES OF THE MOLECULE

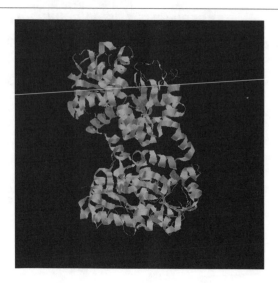

Experiment 1
SKILL BUILDING LAB: SPECTROPHOTOMETRY OF DYES

Pre-Laboratory Assignment   **Due Before Lab Begins**

NAME: _____

Complete these exercises after reading the experiment but before coming to the laboratory to do it.

1. Refer to Figures I-1 and I-2 in the Introduction, which depict a "color wheel" and the visible spectrum, and use these to determine

    a. The color complementary to violet.

    b. The color corresponding to a wavelength of 450 nm.

    c. The color if all light with wavelengths above 600 nm and below 530 nm were absorbed by a sample.

2. The absorption spectrum of chlorophyll *a* has maxima at 663 and 420 nm. What colors correspond to these wavelengths? What color will be seen reflected by an object with this absorption spectrum?

3. a. A sample absorbs 80% of the light passing through it at a certain wavelength. Determine the absorbance of this sample at that wavelength.

    b. Determine the absorbance at the same wavelength if the sample is diluted by half and examined in a cell of equal length.

4. What is one important safety precaution for this experiment?

Experiment 1
SKILL BUILDING LAB: SPECTROPHOTOMETRY OF DYES

BACKGROUND

In this experiment you will be introduced to the use of UV–visible spectrophotometers, which measure the light absorbed in the ultraviolet and the visible regions of the electromagnetic spectrum.

One type of UV–visible spectrophotometer is the *scanning* spectrophotometer. In it the grating or prism is physically moved to cause different wavelengths to fall on the single detector. An advantage of scanning is that the optics can be tuned to give high wavelength resolution. However, any scanning system must always include at least one moving part, and, aside from problems of breakage, recording a spectrum may take a considerable length of time because the whole spectrum is sampled one wavelength at a time. The Spectronic-20 spectrophotometers used in many undergraduate labs and in some clinical settings could be used as scanning instruments if you set each new wavelength manually.

The second type of UV–visible spectrophotometer is the *diode-array* spectrophotometer. In it a narrow beam of white light is sent through a sample and then refracted through a prism onto a grid (or array) of hundreds of detectors based on diodes. The detectors are physically arranged so that each position receives only a narrow range of wavelengths (in some instruments 2 nm wide). All of the diodes detect the light simultaneously, so there is a significant decrease in the amount of time it takes to sample many wavelengths. There is only one simple moving part: a shutter. However, the resolution is fixed by the density of the array and can never be improved. The detection of hundreds of different signals requires fast processing and data storage—precisely the strengths of a computer-controlled instrument.

 Caution: The dyes in this experiment are related to common food dyes, but you will also be working with concentrated acids and bases. Even when using the spectrophotometers, careful handling and safety goggles are essential.

PROCEDURE

There are three parts to this lab. You will prepare a solution of a food dye and obtain its absorbance spectrum using the spectrophotometer. Next, you will combine your solution with that of another student to get a mixture of dyes, which you will also analyze on the spectrophotometer. Finally, you will make derivatives of your dye in a solution of acid and a solution of base.

Carefully review the directions for the operation of the spectrophotometer in your laboratory. You will need to know how to obtain and print spectra.

For all of your solutions, including those obtained by dilution, make careful notes of the color and intensity.

Part I: Recording of Standard Spectrum (Individual Work)

When you come to lab, your instructor will assign you a red, yellow, or blue dye. Prepare a stock solution by dissolving four drops of the dye in water in a graduated cylinder to make a 100-mL solution. Mix thoroughly.

When you are ready, use a Mohr pipet to transfer 3.00 mL of your stock solution to an 18 × 150-mm test tube. Add 7.00 mL of water and mix well.

Prepare a second sample from 2.00 mL of your stock solution and 8.00 mL of water.

Prepare a third sample from 1.00 mL of your stock solution and 9.00 mL of water.

Then transfer a portion of each sample to a different cuvette. Take them to the spectrophotometer, obtain an absorption spectrum, and plot the spectrum.

Part II: Preparation of Mixtures (Paired Work)

You should see that your spectra have a single prominent absorption. Other students with different dyes will have samples that absorb at different wavelengths. Ask one other student to share his or her solution with you. Prepare a solution in an 18 × 150-mm test tube using 3.00 mL of your stock solution, 3.00 mL of the other person's stock solution, and 4.00 mL of water. Make sure the solution is well mixed. Then transfer a portion to a cuvette and record its spectrum.

Part III: Preparation of Derivatives (Paired Work)

Dyes are very sensitive to their chemical environment. Clean out two 18 × 150-mm test tubes and label them 1 and 2. Put 3.00 mL of your stock solution and 4.00 mL of water in each.

To test tube 1 add 3.00 mL of 6 M HCl. Mix the solution well. Note any color change and how long it takes. Transfer a portion to a cuvette.

To test tube 2 add 3.00 mL of 6 M NaOH. Mix the solution well. Note any color change and how long it takes. Transfer a portion to a cuvette.

Obtain an absorbance spectrum for each of the solutions.

When you are done, return the samples to the appropriate test tubes. Take test tube 1 and add 3.00 mL of 6 M NaOH to it. Note any color changes. Take test tube 2 and add 3.00 mL of 6 M HCl to it. Note any color changes. Do not take spectra of these solutions.

REPORT

Your report should show your understanding of the principles of spectrophotometry discussed in the Introduction. Tabulate the wavelengths of the peaks in each of the spectra. Determine %T and A values for the peaks.

Attach copies of your spectra to your report.

QUESTIONS TO ANSWER IN YOUR REPORT

1. For each of the spectra you took, what is the relationship of the absorbed wavelength and the color you see with your eyes?
2. Your dye should obey Beer's Law. Do your data support this hypothesis?
3. What happened when you combined your solution with that of another student? Was the intensity of your dye's absorption altered by the presence of the second dye?
4. What changes did or did not occur when you treated your solution with acid and with base? Do not just comment on the perceived color; analyze your spectra also.

Experiment 2
FOUNDATION LAB: SPECTROPHOTOMETRIC DETERMINATION OF Cu^{2+} ION

Pre-Laboratory Assignment   **Due Before Lab Begins**

NAME: _____

Complete these exercises after reading the experiment but before coming to the laboratory to do it.

1. Describe how to dilute a 0.300 M Cu^{2+} solution to make 25.00 mL of solutions that are 0.100 M, 0.150 M, and 0.200 M Cu^{2+}.

2. Define "blank" and describe what blanks you will use for each of your measurements.

3. If you have 0.40 mL of 0.200 M Cu^{2+}, how many mL of 0.50 M NH_3 must be added so that the ratio of moles of NH_3 to moles of Cu^{2+} is 10:1?

4. What safety precautions are required when handling the ammonia solution? Why?

Experiment 2
FOUNDATION LAB: SPECTROPHOTOMETRIC DETERMINATION OF Cu^{2+}

BACKGROUND

Solutions containing copper(II) ions (Cu^{2+}) have a characteristic pale blue color, assuming the anion present is colorless and no other colored substances are present. The intensity of the color is dependent upon the concentration of the copper(II) ions present. When aqueous ammonia, NH_3 (*aq*), is added to the solution, a different dark blue color develops as the $[Cu(NH_3)_4]^{2+}$ complex ion forms. The ammonia is known as a derivatizing agent.

$$Cu^{2+}\ (aq) + 4\ NH_3\ (aq) \rightarrow [Cu(NH_3)_4]^{2+}\ (aq)$$

You and your group will prepare a series of solutions of known concentration in order to make one of two calibration curves, either absorbance versus concentration of Cu^{2+} or absorbance versus concentration of $[Cu(NH_3)_4]^{2+}$. You decide which solutions will give a more appropriate calibration curve and use the curve to determine the concentration of Cu^{2+} in an unknown.

The $[Cu(NH_3)_4]^{2+}$ system may deviate somewhat from Beer's Law because of the tendency of the $[Cu(NH_3)_4]^{2+}$ ion to react with water to form $[Cu(NH_3)_3H_2O]^{2+}$ and other ions that do not absorb at the same wavelength maximum as the $[Cu(NH_3)_4]^{2+}$ ion. This is an indication that the formation of the $[Cu(NH_3)_4]^{2+}$ ion is an equilibrium reaction. This must be "driven" by the addition of excess ammonia. Even in that case, you must use your calibration curve instead of Beer's Law in your analyses of the unknown.

Caution: The ammonia solution used in this experiment is a potentially toxic, corrosive solution. Do not inhale the ammonia fumes directly because these can irritate the respiratory system. As usual, be sure to wear appropriate eye protection at all times in the lab and wash your hands before leaving the lab.

PROCEDURE

Part I: Assign Tasks to Group Members

You will be working in a group of three to four people. Everyone is expected to participate in the experiment, to record data in his or her own notebook, and to complete his or her own report. Refer to the Introduction for more information on how to form and work in groups.

Part II: Determination of the Appropriate Solution to Use to Analyze for Cu^{2+} (Group Work)

Most standard cuvettes can hold 3 mL of solution. Determine the volume of the cuvettes you will use in the experiment and adjust the volumes given in this procedure if the volume of the cuvette is different from 3 mL.

Fill one cuvette with 3 mL of stock Cu^{2+} solution. The stock solution is the solution of known concentration from which dilute solutions will be prepared. Prepare another cuvette containing appropriate volumes of Cu^{2+} solution and ammonia solution, the derivatizing agent, so that the total volume in the cuvette does not exceed

3 mL. As a group, you need to decide upon "appropriate volumes." Choose a volume of Cu^{2+} solution and calculate the volume of NH_3 solution needed so that you have 8 to 10 times as many moles of ammonia as Cu^{2+}. If the sum of the volumes exceeds 3.00 mL, try a different volume. However, try not to just guess, but reason your way to a different volume based upon the results of your first calculation.

Transfer, using pipets, the agreed-upon volumes of Cu^{2+} and NH_3 solutions to a cuvette. Cap the cuvette and mix well. Mixing measured amounts of two solutions to make a new solution with a total volume of 3.00 mL assumes that the volumes are additive. Although this is not always the case, we have verified that it is a reasonable assumption in this experiment.

You need to prepare two blanks to calibrate the spectrophotometer: one for use with the Cu^{2+} solution and one for use with the $[Cu(NH_3)_4]^{2+}$ solution. A blank is a solution that contains all of the substances present in the samples being analyzed except the analyte (the substance of unknown concentration). Decide upon appropriate blanks and then check with your instructor before you prepare the blanks in cuvettes.

Measure the visible spectrum of each of the two solutions (the Cu^{2+} solution and the $[Cu(NH_3)_4]^{2+}$ solution). If your spectrophotometer is capable of scanning the entire visible range, scan the entire visible range so that you can determine the wavelength of maximum absorbance for each of your solutions. If your spectrophotometer lacks that feature, see your instructor in order to determine the wavelength at which to set your instrument.

Analyze the results of the two spectra to determine whether to use Cu^{2+} (aq) alone or $[Cu(NH_3)_4]^{2+}$ (aq) in preparing the calibration curve and analyzing the solution of unknown Cu^{2+} concentration. You can also determine whether or not you may use a solution as concentrated as the stock solution when preparing your calibration curve. Remember, the absorbance range for analytical purposes is between 0.1 and 1.5. Discuss with your group members whether the stock solution is an acceptable concentration. If not, then look at the absorbance and make your most concentrated solution be the right concentration so that the absorbance is less than 1.5.

Part III: Preparation of the Solutions and the Calibration Curve (Group Work)

Prepare at least four different Cu^{2+} solutions beginning with the stock Cu^{2+} solution. Use a separate volumetric flask for each solution. These solutions are called standard solutions because their concentrations are known. Good analytical procedure requires that all samples be treated in the same manner. This involves four steps:

1. Prepare solutions in volumetric flasks containing known amounts of the species you are studying.
2. Transfer the same amount of each solution to a cuvette.
3. Add the same amount of the derivatizing agent (or pure water, if no derivatizing agent is used) to the cuvette. Cap the cuvette and mix well.
4. Measure the absorption spectrum of the solution in the cuvette.

If your group decided to use the Cu^{2+} solution alone, then pipet 3 mL of each of the standard solutions into separate cuvettes. If your group decided to use the $[Cu(NH_3)_4]^{2+}$ solution, pipet the same amount of each of the standard solutions (use the same volume as the volume of stock solution you used in Part II of this experiment) to separate cuvettes and then add the appropriate volume of aqueous NH_3 (as determined in Part II of this experiment).

Measure the absorbance of each of the calibration standards with the spectrophotometer set at the wavelength of maximum absorbance. Remember, absorbances need to be between 0.1 and 1.5.

Part IV: Determination of Copper(II) Concentration in an Unknown (Individual Work)

Obtain a solution of unknown Cu^{2+} concentration from your instructor. Treat the unknown exactly as you did the standards.

CALCULATIONS

To ensure good results, Parts II through IV should be done during the lab period so that data that do not fit the calibration curve can be remeasured.

1. Plot a calibration curve showing the absorbance as a function of the concentration of either Cu^{2+} or $[Cu(NH_3)_4]^{2+}$ in the standard solutions (choose whichever species was present in your cuvettes). If you used $[Cu(NH_3)_4]^{2+}$ solutions, then to determine the concentration of $[Cu(NH_3)_4]^{2+}$ in each of the standard solutions, assume complete reaction between the Cu^{2+} and the NH_3 with Cu^{2+} as the limiting reactant. Remember, you diluted each of your standard solutions when you prepared the cuvettes.

2. Examine the calibration curve. If any points are grossly out of line from the curve or if any of the measured absorbances are out of the appropriate range ($0.1 < A < 1.5$), adjust the dilutions if necessary and repeat the measurements.

3. Use the calibration curve to determine the concentration of Cu^{2+} in your unknown.

RESULTS AND CONCLUSION

Organize your experimental results. Be sure to include the wavelength used in the experiment, the absorbance of each solution at that wavelength, and the concentration of each solution. Include your calibration curve and sample calculations where appropriate.

In your conclusion, summarize the experimental method used to analyze the unknown. Include a comparison of the absorbance spectra of Cu^{2+} to that of $[Cu(NH_3)_4]^{2+}$. Report the concentration of Cu^{2+} in the solution of unknown concentration. Obtain the expected concentration of Cu^{2+} in the unknown from your instructor and calculate your % recovery.

$$\% \text{ recovery} = (\text{experimental value}/\text{expected value}) \times 100$$

Ideally your % recovery will be 100%. However, some error is always expected whenever measurements are involved. Analytical chemists determine an acceptable range for the results of a given analysis dependent upon the accuracy of the equipment, glassware, and instruments involved in the analysis. If you are within 2% of the expected concentration, your error is within acceptable limits. If not, explain what factors may have contributed to your error.

QUESTIONS TO ANSWER IN YOUR REPORT

1. Was the method you used to determine the concentration of Cu^{2+} in the unknown effective? Explain.

2. If different members of your group analyzed the same unknown, compare your results and account for any differences in the reported concentration of the unknown.

3. Industries, such as environmental laboratories, use spectrophotometric methods to determine the concentration of different substances in solution. They typically ignore the dilution factor when the analyte solution is added to the cuvette along with other derivatizing agents *if they treat the unknown in exactly the same manner as the standards.* For example, in this lab, you can ignore the dilution factor introduced when you prepared the cuvettes if you used the same volume of Cu^{2+} solution and the same volume of NH_3 solution in each cuvette. Explain why the dilution factor can be ignored under these conditions.

Experiment 3
APPLICATION LAB: TOTAL SERUM IRON ASSAY

Pre-Laboratory Assignment   **Due Before Lab Begins**

NAME: _____

Complete these exercises after reading the experiment but before coming to the laboratory to do it.

1. Since you will be using spectrophotometry to determine the concentration of iron in solutions, what would be useful for you to generate first to accomplish this task? (Recall the analysis of Cu^{2+} experiment.)

2. What is the appropriate absorbance range for spectrophotometry?

3. a. When using FerroZine to determine the concentration of iron, how many moles of FerroZine are needed to complex 1.8×10^{-7} mol of iron?

   b. How many mL of 100 mg/dL FerroZine solution are needed to obtain the number of moles of FerroZine determined in part (a)?

4. a. How many moles of NH_2OH are present in 0.50 mL of 0.434 M NH_2OH?

   b. How many moles of Fe^{3+} can be reduced by the NH_2OH you calculated in part (a)?

5. As part of this laboratory, you may have to remove iron from a solution. If you have 20 mL of a solution that contains 43 $\mu g\ dL^{-1}$, how many grams of iron must be removed to reach a concentration of 23 $\mu g\ dL^{-1}$; how many moles of iron is this?

6. As part of this laboratory, you may have to add iron to a solution. If you have 20 mL of a solution that contains 15 μg dL^{-1}, how many grams of iron must be added to reach a concentration of 23 μg dL^{-1}; how many moles of iron is this?

7. What special precautions must you take while handling the hydroxylamine and deferoxamine solutions? Why?

Experiment 3
APPLICATION LAB: TOTAL SERUM IRON ASSAY

BACKGROUND

The analytical method for this lab is based on a reagent, FerroZine, that forms an intensely colored solution with iron. FerroZine complexes with Fe^{2+} forming a magenta colored solution whose intensity is dependent upon the concentration of the complex present. The equation for the reaction, written for the dianion of FerroZine, is:

$$[Fe(H_2O)_6]^{2+} + 3 \text{ FerroZine}^{2-} \rightarrow Fe(\text{FerroZine})_3{}^{4-} + 6 H_2O$$

The FerroZine is a water soluble salt (with a molar mass of 492.47 g/mol) which forms a -2 anion in solution. Since FerroZine binds Fe^{2+}, all of the iron in solution must be present in the $+2$ state before the concentration is measured. The Fe^{3+} is readily reduced to Fe^{2+} by reaction with excess hydroxylamine:

$$2 Fe^{3+} + 2 NH_2OH \rightarrow 2 Fe^{2+} + N_2 + 2 H^+ + 2 H_2O$$

Thus, to prepare the spectrophotometric solutions, put a fixed amount of standard solution into the cuvette along with a fixed amount of hydroxylamine and FerroZine. This means you will follow a two-step procedure to prepare the spectrophotometric solutions. First, dilute the iron stock solution in a volumetric flask. Then, put a fixed amount into a cuvette, along with a fixed amount of hydroxylamine and FerroZine.

In the foundation, you needed to be sure that you added enough ammonia to complex all the copper(II) ions. Here, you need to be sure that there is enough hydroxylamine and FerroZine in each cuvette to reduce and complex all of the iron. Therefore, start off by calculating the largest concentration of iron you might have, and then determine how many moles of FerroZine and of hydroxylamine you might need. This will tell you how to make up the cuvettes for the spectrophotometer.

The measurement $\mu g/dL$ (micrograms per deciliter) is a common concentration unit used in medical analyses. You need to be able to convert between molarity and this unit in order to report your results in the units used in a clinical setting.

Caution: You will not be working with real blood samples. Your samples will be solutions of iron(II) ammonium sulfate, $Fe(NH_4)_2(SO_4)_2$. If you were using real blood samples, you would have to observe the necessary precautions for the safe handling of a blood sample. These include wearing protective gloves, disposal of the glassware used to contain the blood sample and the latex gloves in a biohazard bag after analysis is complete, and washing any spills with bleach.

This lab requires the use of a hydroxylamine solution. It is potentially toxic and a potential mutagen. This lab also involves the use of deferoxamine, a biologically active iron-complexing agent and possible teratogen. The use of proper gloves is advised. Wear appropriate safety goggles. Clean up all spills immediately with excess water. Be careful to avoid ingestion and, as always, wash your hands carefully before leaving the lab.

PROCEDURE

This procedure will be similar to the one you used in the foundation lab, except for the "treatment" step at the end.

You will have the following materials at your disposal for this lab:

- iron(II) standard solution
- FerroZine solution
- hydroxylamine solution
- solutions containing "normal" levels for iron
- spectrophotometer
- appropriate glassware
- deferoxamine, an iron complexing agent (shown in Figure B-2).

Part I: Formation of Groups

The group in this experiment will emulate a team working in a clinical analytical lab. Three students will share the responsibility for constructing a calibration curve based upon the serum iron levels in "healthy" patients.

Part II: Determining the Response of Healthy Iron Levels in the Assay (Group Work)

The standards will be available in the lab in the form of nine "healthy" patient samples. Look on the bottles of "sample sera" available in the lab to get the actual iron concentrations.

These nine samples will form your calibration curve for the experiment. Since your unknown patient sample contains either too much iron or too little iron, compared to the acceptable "normal" range for that patient, your calibration curve needs to extend beyond the normal range at both ends. Therefore, you need to prepare two more samples in addition to the nine "healthy patient" samples; one solution that is more concentrated than the highest serum iron level in a healthy patient and one that is more dilute than the lowest serum level in a healthy patient. Use the standard iron(II) solution available to prepare these two solutions in volumetric flasks.

The easiest way to set up the assay is to determine how to take the most concentrated sample and to mix it in a cuvette to give a reasonable A in the spectrophotometer, with appropriate amounts of FerroZine and hydroxylamine. Treatment of the other samples with identical dilution methods will fill out the calibration curve.

Part III: Treatment of Unhealthy Iron Levels

A patient with too much or too little iron must be treated with iron supplements or iron-removal agents. You will simulate this during the lab by adding a supplement or a removal agent to the serum of your patient. Get your patient's serum, and use the method of your group to determine the iron concentration. Then, adjust the amount of iron in the serum to bring it to the normal range.

FIGURE B-2 MOLECULAR STRUCTURE OF DEFEROXIMINE, $C_{25}H_{48}N_6O_8$

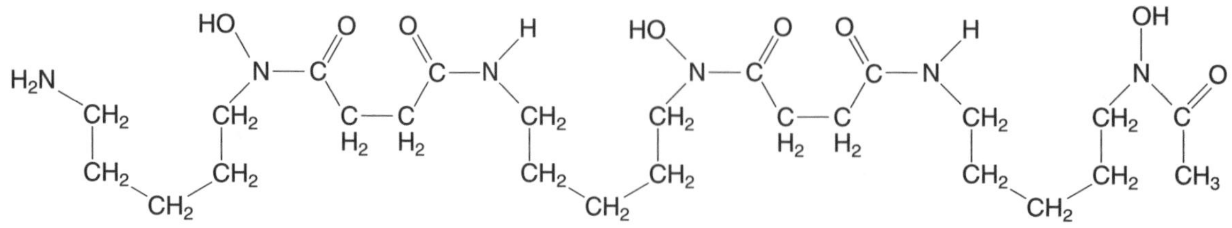

In a clinical setting, compounds that contain iron(II) are used in the treatment of iron deficiency anemia, which is the condition of too little iron in the blood. If your sample is "anemic," you will mix in a solution of iron(II) sulfate. Calculate how many moles of iron you need to add to the solution, then add the appropriate volume of iron(II) solution to add that amount of iron to your sample. Note that this will dilute your solution by a small amount, but you are aiming for a solution that, when you determine the iron content, is in the normal range for your patient. After mixing in the iron supplement solution, withdraw some of the solution and redetermine the iron concentration.

If you needed to *remove* iron in a clinical setting, you would administer an agent that would bind to the iron and cause it to be excreted. We simulate this by using the agent to bind to some of the iron and to allow the FerroZine to complex less of the iron in the solution—which will show up as a lower iron concentration. Deferoxamine is used in the clinical treatment of acute iron overdoses. Deferoxamine has a molar mass of 656.8 g/mol and binds to iron(III) ions in a one-to-one mole ratio.

To "remove" iron from your solution, you will mix in a solution of deferoximine. Calculate how many moles of iron you need to remove from the solution, then add the same number of moles of deferoximine. Note that this will dilute your solution by a small amount, but you are aiming for a solution that, when you determine the iron content, is in the normal range for your patient. After mixing in the deferoxamine solution, withdraw some of the solution and redetermime the iron concentration.

RESULTS AND CONCLUSIONS

Organize your experimental results in an appropriate manner. Include graphs if any were generated.

Your conclusion should include a summary of the experimental technique you and your group developed in order to measure the serum iron level in the patient's blood. What are normal serum levels for adult males, females, and children? What were the results of your patient's sample? How did you treat the patient? Was the treatment successful? Explain how you know.

QUESTIONS TO ANSWER IN YOUR REPORT

1. Iron(II) sulfate and iron(II) fumarate are often used in iron supplements for the treatment of iron deficiency anemia. Why do you think these compounds are used?
2. The body only retains 10 to 15% of the iron consumed. A prenatal multivitamin contains 45 mg of iron per tablet. Assuming maximum retention (15%) and a volume of serum of 8.0 L, determine the concentration, in micrograms per deciliter, of iron after ingestion of the vitamin.

EXPERIMENT GROUP C

ANALYSIS OF COMPLEX SOLUTIONS: BLOODWORK QUICK WITH A "FINGER STICK"

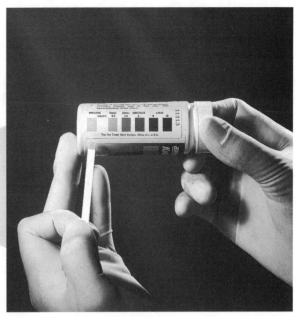

Measurement with a test strip is an important way to quickly assess a patient's health.
Saturn Stills/Science Photo Library/Photo Researchers.

PURPOSE Chemists often use the way that light interacts with matter, also known as spectrophotometry, to determine how chemical substances interact with each other. This is especially useful in studying solutions, including solutions in which two or more substances form a complex with each other. This experiment module will develop your skills in analyzing a solution using UV and visible light, and will show you how the analysis of a complex of a dye with a blood protein permits the determination of important health-related parameters.

SCHEDULE OF THE LABS

EXPERIMENT 1: Skill Building Lab: "Stoichiometry of a Metal–Ligand Complex"

1. Microscale determination of molar ratios (group work).
2. Preparation of solutions for mole ratio analysis (individual work).
3. Spectrophotometric determination of the formula of a complex (group work).

EXPERIMENT 2: Foundation Lab: "K_a of an Indicator"

1. Determining the pH transition range for a series of indicators (group work).
2. Microscale determination of the approximate range of pK_a of the indicator (individual work).
3. Spectrophotometric determination of the approximate range of pK_a of the indicator (individual work).

EXPERIMENT 3: Application Lab: "Determination of Serum Albumin"

1. Spectrophotometric study of the binding of bromocresol green and albumin (group work).

67

2. Design of a finger-stick" test (group work).
3. Measurement of serum albumin (individual work).

SCENARIO Any organism with a circulatory system has the challenge of forcing blood through tubes known as arteries and veins. The heart applies the pressure needed for blood to flow, generating a mechanical pressure known as the blood pressure. The arteries and veins, especially when very small, have walls that are permeable to water and other small un-charged solutes. But ions and larger molecules cannot pass through the walls of the blood vessels. The blood pressure could, if left unchecked, force most of the water from the blood, much as water can seep through a wall with cracks.

The only mechanism to counteract the blood's mechanical pressure is osmotic pressure, which describes the tendency of water to be retained by a solution. In mammalian blood the osmotic pressure is regulated by a large molecule called albumin. Proper concentrations of albumin are found in healthy individuals, but many diseases can cause albumin to be lost via the urine, affecting the entire circulatory system.

One of the first analyses done when a patient is admitted to a hospital is the determination of the serum (blood) albumin level. This is done on an analyzer that uses the exact same method you will use in this lab. The serum is diluted into a solution containing an acid–base buffer and an excess of the dye bromocresol green (BCG). BCG is normally yellow under these conditions, but it forms an intensely green colored complex with albumin. The green albumin–BCG complex is the basis for this experiment.

Many analytical methods rely on a less precise, but much quicker, method of a simple finger-stick analysis. A drop of blood is placed on a small reagent strip and the chemical substances on the strip turn color, depending on the amount of material in the blood. This can then be compared with a standard color strip to get a rough value for the concentration of the material in the blood. In this lab, you will carry out both finger-stick and more precise determinations of the albumin concentration, simulating the professionals who must obtain quick results prior to sending certain samples to a lab.

Experiment 1
SKILL BUILDING LAB: STOICHIOMETRY OF A METAL–LIGAND COMPLEX

Pre-Laboratory Assignment   **Due Before Lab Begins**

NAME: _____

Complete these exercises after reading the experiment but before coming to the laboratory to do it.

1. A solution is made by dissolving 1.00 g of the ligand bipyridine ($C_{10}H_8N_2$) in enough water to make 1.00 L of solution. Determine the concentration of bipyridine in the following solutions:

   a. Take 10.0 mL of the original solution and add enough water to make 100. mL of solution.

   b. Take 0.40 mL of the original solution and mix in a cuvette with 0.50 mL of hydroxylamine and 1.40 mL of iron solution.

   c. Mix 1.00 mL of the solution from part (a) with 1.20 mL of hydroxylamine solution and 0.40 mL of iron solution.

2. In this lab, you will also carry out dilutions using a dropper to get qualitative information about metal/ligand ratios. Determine the approximate answers to these questions:

   a. Put two drops of hydroxylamine solution in a 0.50-mL well. Then add eight drops of 1.00 g L^{-1} bipyridyl and two drops of 5 mg L^{-1} iron. How many moles of bipyridine and how many moles of iron are in the well? There are about 20 drops to 1 mL.

   b. If you want to repeat the mixture from part (a) on a larger scale, describe how many milliliters of hydroxylamine solution, how many milliliters of bipyridine solution, and how many milliliters of iron solution you need to make 2.50 mL of total solution volume.

3. List two different safety precautions that are important in this laboratory.

Experiment 1
SKILL BUILDING LAB: STOICHIOMETRY OF A METAL–LIGAND COMPLEX

BACKGROUND

A study of the absorption of electromagnetic radiation by a sample is one of the best ways chemists have to determine the chemical contents of a solution without an extensive process of reaction analysis that usually causes the destruction of the sample. The study of interaction of radiation with matter is the field of spectrophotometry. When the radiation is in the ultraviolet and the visible regions of the electromagnetic spectrum, this is called UV–visible spectrophotometry.

UV–visible spectrophotometry is also called electronic spectrophotometry, because absorption of UV and visible light often causes a redistribution of the electrons in the absorbing species. The electromagnetic energy of light causes the electrons to move to one or more excited states. Later on, the electrons return to their original configuration, also known as their ground state. Sometimes, however, the electronic excited state gives rise to chemical reactions, an area called photochemistry. The synthesis of new chemical substances using light energy is the process we call photosynthesis.

In this experiment group, you study the nondestructive examination of a solution to determine the chemical contents. You examine how spectrophotometry can be used to answer the question, "How much of a certain substance is in the solution?" You will also work on the problem of determining *what* is in solution.

DETERMINATION OF MOLE RATIOS IN COMPLEXATION: A GRAPHICAL METHOD*

If any two substances A and B are mixed in solution, they may in principle react to make one or more new substances. In many cases this is a simple addition reaction in which one or more moles of substance B add to a mole of A to give a complex:

$$A + nB \rightarrow AB_n$$

In general such reactions do not proceed to completion. A mixture of A, B, and AB_n are then left in equilibrium. Equilibrium is an important phenomenon and must often be dealt with. Fortunately, in this experiment group the reactions can be assumed to go to completion. We can also assume that only one AB_n complex forms, with nB for every A.

The species AB_n is referred to as a complex of A and B. A common type of complexation occurs when A is a metal ion that reacts with other small molecules or ions (B) to form a metal coordination complex. In those cases, the term "coordination" refers to the connection of B and A, so the A—B bond is sometimes called a coordinate bond. In such a complex, B is referred to as a ligand for A.

The amount of AB_n formed depends on the amounts of A and B in the mixture and on the value of n. Let's say that n is 4. This is the case, for example, in the formation of the complex $[Cu(NH_3)_4]^{2+}$ when a solution of a copper(II) salt is mixed with concentrated ammonia solution:

---

* This method follows that outlined in Po, H. N., and Huang, K. S.-C. *J. Chem. Educ.*, **1995**, *72*, 62–63.

71

$$Cu^{2+}(aq) + 4\ NH_3\ (aq) \rightarrow [Cu(NH_3)_4]^{2+}\ (aq)$$

In this case, ammonia acts as a ligand.

If we prepare a solution that contains 0.0020 mol of Cu^{2+}, the amount of complex that forms depends on which reactant is the limiting reactant. If there is less than 0.0080 mol of ammonia in the solution, then the ammonia is the limiting reactant and, regardless of the amount of copper, we will only have as much complex as the supply of ammonia permits.

If we have more than 0.0080 mol of ammonia, copper is the limiting reactant. Then we would form 0.0020 mol of $[Cu(NH_3)_4]^{2+}$ complex. The copper should be the limiting reactant if, for example, we want to use the characteristic deep blue color of $[Cu(NH_3)_4]^{2+}$ to determine the amount of Cu^{2+} in a solution.

You may be familiar with using spectrophotometry to determine the amount of a substance in a solution. But did you ever wonder where the recipe for mixing comes from? It requires that someone carefully determine the relative values of the two reactants, so that every solution that is analyzed has enough ligand. The point of today's exercise is to for you to determine n for a series of ligands that complex with Fe^{2+}.

Two Important Points

Protection of Iron(II) from Oxygen

We will be working with the Fe^{2+} ion in this experiment. However, this ion reacts rapidly with air to give Fe^{3+} in acidic or even neutral solutions:

$$4\ Fe^{2+}\ (aq) + O_2\ (g) + 4\ H^+\ (aq) \rightarrow 4\ Fe^{3+}\ (aq) + 2\ H_2O\ (l)$$

Fortunately, if you do not splash, stir, or shake a solution rapidly, then oxygen is slow to get into the solution. To correct for the small amount of Fe^{3+} that forms during normal handling, we always put some of the reducing agent hydroxylamine, NH_2OH, in the solution to keep all the iron in the +2 state.

Determination of a Ligand-to-Metal Ratio

There are two ways to determine the ligand-to-metal ratio in a system where a complex is formed completely upon mixing solutions. One way is to carry out a spectrophotometric titration, in which we keep adding metal to a solution of the ligand until there is no further change in the color. But this approach assumes that the color of the complex dominates what we see and that we can suddenly say "there is no deepening of the color any more." That is a difficult thing to judge, even with an instrument.

The second way to determine n is to carry out the addition of increasing amounts of ligand to solutions that contain less and less metal. In the first solutions, there is not enough ligand: The ligand is the limiting reactant. In the later solutions, there is not enough metal. The procedure gives rise to a "peak" solution, in which the ligand-to-metal ratio is just right. That solution has a distinct color compared to those before and after it in the series. We then find the ratio of ligand to metal that went into this solution.

Table C-1 gives an idea of how the second method works. The ligand-to-metal ratio n is 1:4. We start with 1.0×10^{-6} mol (or 1.0 micromole, 1.0 μmol) of metal, and 0.25×10^{-6} mol (0.25 μmol) of ligand. In each successive solution, we increase the volume of the ligand solution by 0.20 mL and decrease the volume of the metal solution by the same amount. As a result, the number of moles of complex increases, then begins to decrease once we pass the ideal 2:1 ratio. This occurs somewhere between sample C (ratio 1.25:1) and sample D (ratio 2.5:1).

TABLE C-1

Solution	Volume of Metal (0.0010 M)	Moles of Metal	Volume of Ligand (0.00125 M)	Moles of Ligand	Moles of Complex
A	1.00 mL	1.00×10^{-6}	0.20 mL	0.25×10^{-6}	0.12×10^{-6}
B	0.80 mL	0.80×10^{-6}	0.40 mL	0.50×10^{-6}	0.25×10^{-6}
C	0.60 mL	0.60×10^{-6}	0.60 mL	0.75×10^{-6}	0.38×10^{-6}
D	0.40 mL	0.40×10^{-6}	0.80 mL	1.00×10^{-6}	0.40×10^{-6}
E	0.20 mL	0.20×10^{-6}	1.00 mL	1.25×10^{-6}	0.20×10^{-6}

We can demonstrate the second method by a simple microscale experiment with drops. We systematically vary the ligand-to-metal ratio by mixing different numbers of drops of each solution, along with any other necessary solution (e.g., hydroxylamine) in the wells of a tissue culture plate. Observation may show a continuous increase or decrease in color, which means that we have the same reactant in excess in all cases. But if we see a "break" in the color, then we have somewhere in the series a solution with the ideal ligand-to-metal ratio.

To get an exact value for the ratio, we use the spectrophotometer. The instrument should be set to record the "best" wavelength for the complex (consult with your instructor about how to do this). We prepare a series of mixtures that have the same ligand-to-metal ratios that gave the "break" in the color in the microscale test. These mixtures may be prepared in volumetric glassware, followed by transfer to cuvettes, or they can be prepared directly in the cuvettes.

The absorbance values A for the five solutions are now recorded at the "best" wavelength. These results are graphed with milliliters of metal solution on the x-axis and A on the y-axis. A properly done graph will look something like the one in Figure C-1. Lines are drawn as shown along the two legs of the graph, and their intersection is noted. This point lies above the spot on the x-axis where the ligand-to-metal ratio is equal to the correct stoichiometric value. From this volume of metal and the corresponding value of the volume of ligand, we can determine the moles of ligand and the moles of metal in the complex. We have determined n.

FIGURE C-1

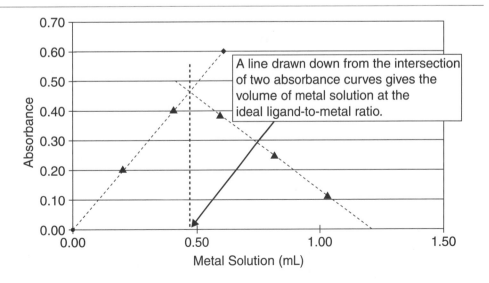

A line drawn down from the intersection of two absorbance curves gives the volume of metal solution at the ideal ligand-to-metal ratio.

⚠ Caution: This laboratory requires the use of hydroxylamine solutions. These are potentially toxic and are potential mutagens. Use of proper gloves is advised. Clean up all spills immediately with excess water.

The ligands have a high affinity for iron. Be careful to avoid ingestion; as always, wash your hands carefully before leaving the lab.

PROCEDURE

There are three parts to this lab. After forming groups and dividing up the ligands, you will use droppers and multiwell tissue culture plates to determine a rough value for the stoichiometric ratio of volumes of the ligand and the metal. Then, you will match this procedure and obtain the spectrum of the relevant complex.

Part I: Formation of Groups

You will work in groups of three today, but each student has essentially a unique set of experiments. Each student works with a different ligand. Some of the ligands will require more dilution than others.

Part II: Microscale Determination of Ligand-to-Metal Ratio

You will have stock solutions for the reaction of iron(II) with three different ligands: FerroZine, terpyridine, and phenanthroline. Before you do any work with the solutions, record the concentrations that are listed on the bottles. Obtain 50 mL of the iron and the ligand and 10 mL of hydroxylamine in separate clean, dry beakers.

All three ligands (Figure C-2) complex Fe^{2+} ion through nitrogen atoms. But one, two, three, or even more of the ligands may complex with the metal at one time. The first step is to determine, qualitatively, how much metal and ligand are needed to provide a good plot. You do this with a tissue culture plate that has wells with a volume of 0.5 mL or less.

FIGURE C-2   LIGANDS FOR ANALYSIS OF IRON

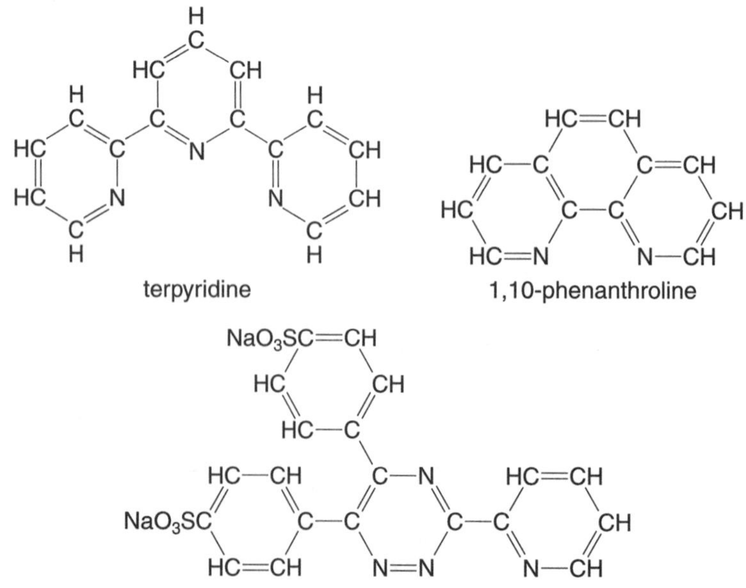

terpyridine

1,10-phenanthroline

FerroZine (sodium salt)

Take the plate and align it so that, from left to right, you have at least five wells. Add three drops of hydroxylamine to each. Then, add 2, 4, 6, 8, and 10 drops of the iron solution to the wells, starting at the left and moving right. Finally, add 10, 8, 6, 4, and 2 drops of the ligand solution, again moving from left to right. Gently swirl the plate to mix the solutions, and put the plate down on a white piece of paper.

Observe the solutions, making careful note of the color and the intensity of the color. Do you see a break in the color? If you do, then you have a range of relative amounts that includes the stoichiometric amount. If you find that the color increases directly with the amount of ligand, then the ligand is the limiting reactant in all cases. You must then dilute the iron. If, on the other hand, the solutions become steadily darker with increasing metal ion, then the iron is the limiting reactant and you need to dilute the ligand.

You have done a range of solutions that cover a 1:5 to 5:1 ratio of volumes of metal to ligand. If it is necessary to dilute the metal or the ligand to bring it into the range for a proper plot, then take the appropriate solution and dilute it by a factor of five. This is done by measuring 20.00 mL of the stock solution into a 100.0-mL volumetric flask, then adding enough water to bring the total solution volume to the mark on the volumetric flask. Do not forget to mix well.

If a dilution was necessary, then after you have the well-mixed diluted solution you should carry out the tissue plate analysis again. This time you should see that you have a well in the center of the range that is the darkest. If not, then try a 10:1 dilution of the original solution.

Part III: Mole Ratio Analysis

Your group now has a solution of the metal and each of the ligands that are appropriate for spectrophotometric analysis. The samples for mole ratio analysis are prepared in the same ratios as the "successful" tissue culture plate experiment. You will have the diluted solutions already (if dilution was needed). Now, prepare five cuvettes with the same ratios of the volume of the metal solution to the ligand solution. In all cases there should be a consistent amount of hydroxylamine in each cuvette, and then a consistent total volume.

This can be planned by filling in the worksheet on the next page.

Wavelengths for maximum absorption of iron(II) complexes

Fe^{2+}–FerroZine	562 nm
Fe^{2+}–Phenanthroline	510 nm
Fe^{2+}–Terpyridine	552 nm

Plotting and Analysis of Spectra

In your group, compare your results and determine if you have the same trends as each other. You may want to work together to convert the volume measurements you have made into the mole amounts you will need for your report.

REPORT

Your report should discuss how the two methods, microscale and spectrophotometric, both gave the information needed to determine the correct ligand-to-metal ratio. Present a figure showing the wells in the tissue culture plate as a function of the number of drops of the metal complex and the ligand (after dilution, if that was needed). Label them carefully by color.

WORSHEET TO GUIDE THE PREPARATION OF SOLUTIONS

IMPORTANT: This is to assist you in planning. All data used in the actual experiment should be recorded in your laboratory notebook.

Part I: Tissue-culture-plate analysis. Fill in the number of drops of each component in the different wells of the tissue culture plate.

	SOLUTION				
	A	**B**	**C**	**D**	**E**
Drops NH_2OH (should be constant)					
Drops Fe^{2+}					
Drops ligand					
Total number of drops (should be constant)					

Part II: Determine the conversion factor between tissue-culture plate and cuvette.

Cuvette volume (mL):	$\dfrac{Cuvette\ volume}{total\ numbers\ of\ drops}$:	mL drop^{-1}

Part III: Spectrophotometric analysis with cuvettes. Multiply each of the values in Part I of the worksheet by the conversion factor in Part II to get the volume in milliliters to use in the cuvette.

	SOLUTION				
	A	**B**	**C**	**D**	**E**
Milliliters of NH_2OH (should be constant)					
Milliliters of Fe^{2+}					
Milliliters of ligand					

For the spectrophotometric experiments, tabulate the volume and concentration of both the metal and the ligand in each of the solutions. Also include a value for A in each of the solutions. A graph presenting volume of metal solution and A, similar to Figure C-1, should be prepared, showing how the two lines converge to the "ideal" value. Finally, calculate the value of n from the ideal mole amounts of the metal and the ligand.

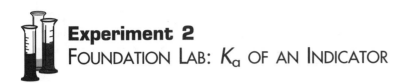

Experiment 2
FOUNDATION LAB: K_a OF AN INDICATOR

Pre-Laboratory Assignment   **Due Before Lab Begins**

NAME: _____

Complete these exercises after reading the experiment but before coming to the laboratory to do it.

    The data in Figure C-3 were collected for the indicator thymol blue. This indicator is yellow at intermediate pH values. It becomes blue at high pH. The absorbance spectra of a series of solutions of thymol blue at different pH values is shown in the figure.

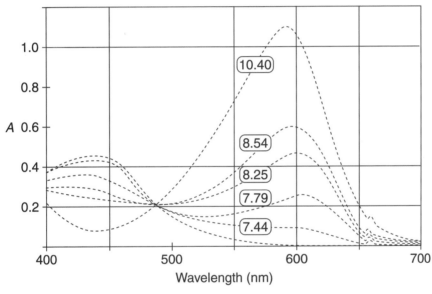

FIGURE C-3

1. For the four intermediate pH's in the graph, determine the value of the fractional ionization α of HA at the pH's listed. Use these values to determine four independent values for K_a. Present your data by filling in the following table.

pH	A_{610}	α	K_a
7.44			
7.79			
8.25			
8.54			
10.40 ("pure base")			

2. Prepare a graph of the absorbance vs. pH of thymol blue at 610 nm in the range from pH = 7 to pH = 8.5. Indicate the approximate point where the absorbance is half of the value at pH = 10.40.

3. What particular safety measures must you take in this laboratory to avoid a false sense of security?

Experiment 2
FOUNDATION LAB: K_a OF AN INDICATOR

BACKGROUND

Spectrophotometry was used in the last experiment to determine the ratio of a ligand to a metal in a complex that, we assumed, was formed completely when two solutions were mixed. This allowed us to use a strategy of varying the reagents to get the ideal ratio.

Other reactions do not proceed to completion when solutions are mixed. Instead, a ratio of species exists in a chemical equilibrium. In these cases, spectrophotometry is still very important to determine the relative amounts of different substances. This experiment uses the same experimental strategy of tissue culture plates and spectrophotometry to deduce the ratio of different forms of a compound that is affected by the acid and base properties of a solution.

An important equation describing the action of an acid in water relates the concentrations of hydrogen ion, H^+, an acid HA, and the acid's conjugate base A^- to K_a, a constant:

$$K_a = \frac{[H^+][A^-]}{[HA]}$$

This relationship holds at all times for a system at equilibrium. The concentration of H^+ may be established through the reaction of the acid HA or the base A^- with water, or the concentration of H^+ could be set through some chemical system so dominant that the concentrations of A^- and HA are forced to follow along. In fact, in research and technological applications it is often more important to ask the question "How does $[H^+]$ influence this chemical substance?" instead of "How does this chemical substance influence $[H^+]$?"

Acid–base indicators are prominent examples of substances that have little influence on pH but whose properties, in their case color, are influenced by $[H^+]$. Their color makes them good candidates for the determination of their acid–base properties using visible light and a spectrophotometer.

Indicators

Indicators are weak acids or weak bases that undergo one or more ionization steps in the range between pH = 0 and pH = 14. All weak acids have a range of pH's for which there is a significant amount of the substance in the acid form (HInd) and a significant amount in the base form (Ind$^-$). An indicator will undergo a sensible color change in this range. In the middle of the transition, the mixture of the ionized and un-ionized forms of the indicator will give a mixed color.*

The color of an indicator can suggest how much of it is in the base form and how much is in the acid form. For example, the indicator bromocresol green is yellow in acid solution and bright blue in basic solution. However, between a pH of about 3.8

---

* The sharp, dramatic color change we need when an indicator is used in a titration is caused by the sudden change in pH from a value well below to a value well above the transition range. This can happen with a fraction of a drop of titrant. The pH changes in this experiment are much more gradual to permit study of the intermediate colors.

and about 5.2 both the acid form and the base form are present. Therefore, the eye perceives the color that is a mixture of blue and yellow—which is green.

The equilibrium for an indicator in solution of controlled pH can be described by the chemical equilibrium equation:

$$HInd\ (aq) \rightleftarrows Ind^-\ (aq) + H^+\ (aq)$$

This suggests a relationship among the concentrations of Ind^-, $HInd$, and H^+, as follows:

$$K_a = \frac{[Ind^-]}{[HInd]}\ [H^+]$$

It is easy to determine the concentration of the hydrogen ion, $[H^+]$ by using a pH meter. However, we also need information about $[Ind^-]$ and $[HInd]$. The "trick" of this lab lies in the fact that we don't need to know what these are. We just need to know the *ratio* of $[Ind^-]$ to $[HInd]$. This is quite easy to do, if all solutions that we study have the *same* total concentration of indicator, [total indicator] = $[Ind^-]$ + $[HInd]$.

Let's say a fraction α of the indicator is in the base form. Then the concentration of the base form will be $[Ind^-] = \alpha$[total indicator]. The fraction of the indicator in the acid form will be $1 - \alpha$, so: $[HInd] = (1 - \alpha)$[total indicator].

That means that, in the equilibrium expression given,

$$K_a = \frac{\alpha[\text{total indicator}]}{(1 - \alpha)[\text{total indicator}]}\ [H^+] = \frac{\alpha}{1 - \alpha}\ [H^+]$$

Absorbance Spectra and Fractional Ionization

Figure C-4 shows the changes in the absorbance spectrum for the indicator phenolphthalein at relevant pH values, and Figure C-5 shows the acid–base equilibrium for phenolphthalein. All the solutions have the identical concentration of phenolphthalein. We see that below pH = 7.97, very little visible light is absorbed by the indicator (here, $\alpha = 0$). At pH 11.38, there is a large amount of light absorbed for the

FIGURE C-4 Changes in the Absorbance Spectrum of Phenolphthalein at Different pH Values

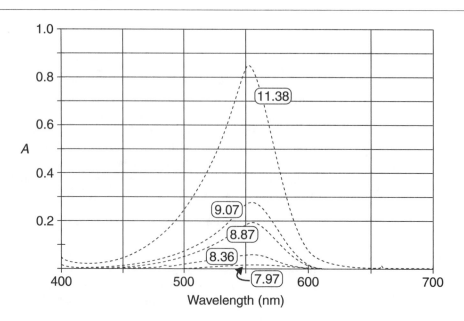

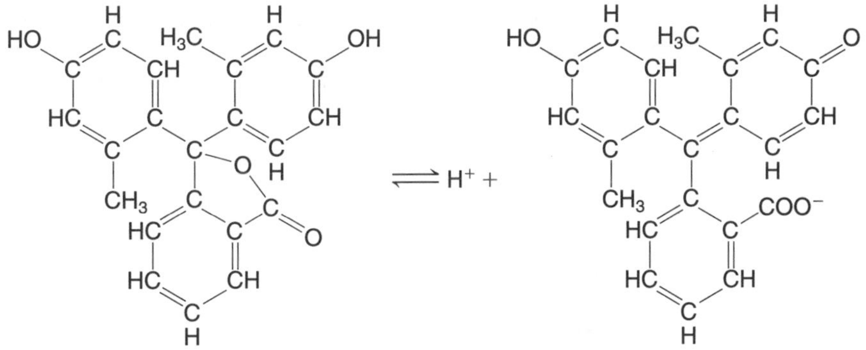

FIGURE C-5 ACID–BASE EQUILIBRIUM FOR PHENOLPHTHALEIN

wavelength 554 nm (here, $\alpha = 1$). In between, the amount of light that is absorbed changes because different amounts of the indicator are in the form that absorbs light at 554 nm. Assuming that phenolphthalein obeys Beer's Law, then the light absorbed is directly proportional to the concentration of the phenolphthalein in the *base* form.

High pH: $A^{554} = A^{554}_{Ind^-/pure}$ $\alpha = 1$

Low pH: $A^{554} = 0$ $\alpha = 0$

Intermediate pH: $A^{554} = \alpha \times A^{554}_{Ind^-/pure}$

$$\frac{A^{554}}{A^{554}_{Ind^-/pure}} = \alpha$$

So, we now have a way to get the fraction ionized, α. We just compare the absorbance at some intermediate pH with that at high pH.*

Let's look for example at the phenolphthalein data. We can determine that the absorbance at pH = 11.38 is 0.842. We can compare the absorbance at other values to get the value for α at those pH:

pH	A^{554}	α
7.97	0.016	0.019
8.36	0.055	0.065
8.87	0.198	0.235
9.07	0.277	0.329

This can now be used to get values for K_a at different values:

pH	α	$[H^+]$	K_a
7.97	0.019	1.1×10^{-8}	2.1×10^{-10}
8.36	0.065	4.4×10^{-9}	3.1×10^{-10}
8.87	0.235	1.3×10^{-9}	4.0×10^{-10}
9.07	0.329	8.5×10^{-10}	4.2×10^{-10}

From this we can calculate an average value of 3.4×10^{-10}. Note that this result may seem imprecise, but it is typical of results for determining an equilibrium constant under general laboratory conditions.

---

* There is an important restriction here, which is met by all indicators you will study in this lab. *Only the base form* of the indicator must absorb light at the wavelength under study. In addition, you must make sure, through good dilution technique, that all solutions have the same total concentration of indicator.

Caution: In this laboratory you handle a variety of different colored substances. The pleasant colors that result make it easy to forget that they are in solutions with significant, and dangerous, acid and base properties. Be certain to wear proper eye protection at all times.

PROCEDURE

Your eye is a sensitive color detector, although it cannot provide very good quantitative data. You will begin this experiment by using tissue culture plates to determine the values of pH where your assigned indicator undergoes a color change. These will allow you to plan for similar experiments in the spectrophotometer.

You will prepare a series of six solutions for analysis. These will include pH's well above and well below the pK_a for the indicator. You will also measure the spectrum for four solutions at pH's where α varies most significantly—within one pH unit of the pK_a.

Part I: Formation of Groups

A group of three students is required for this lab. Each student will have a different indicator—bromocresol purple, bromothymol blue, or thymolphthalein. The structures of these indicators in their *acid form* are given in Figure C-6.

The group should assemble the materials needed for the experiment. One student should be responsible for the tissue culture plates and spectrophotometer cells.

FIGURE C-6

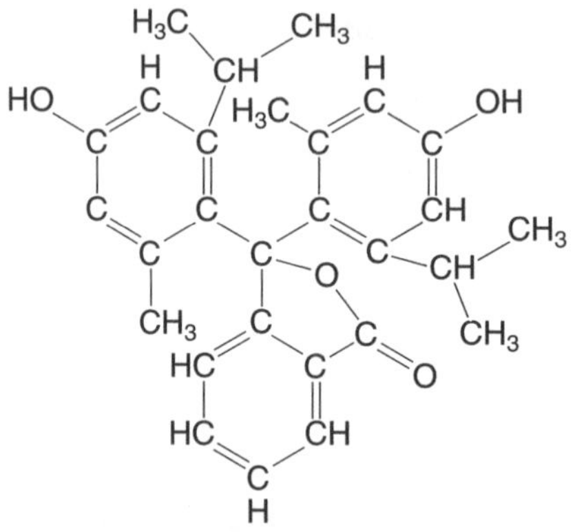

thymolphthalein

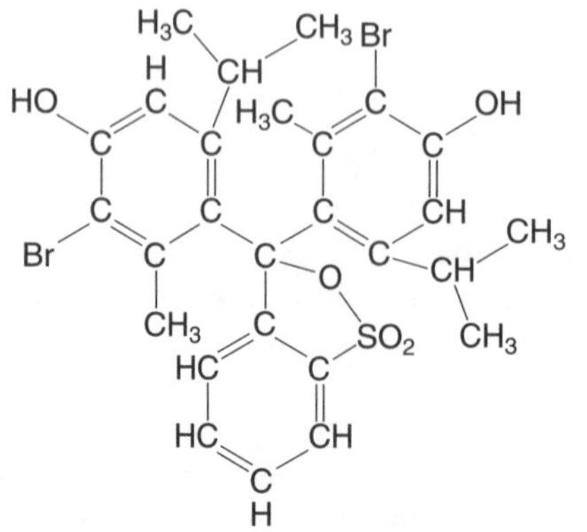

bromothymol blue

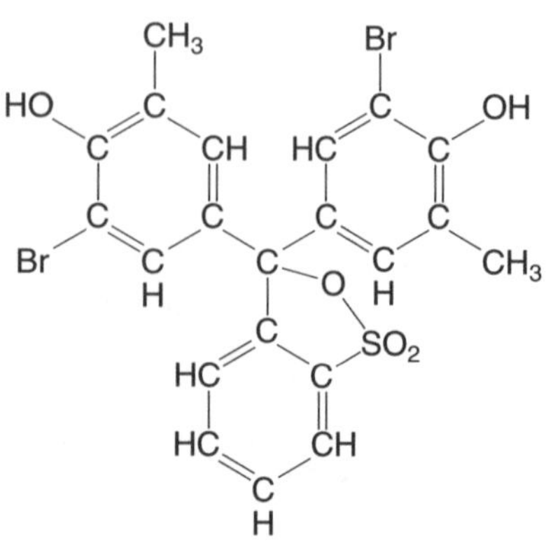

bromocresol purple

The second student should obtain 300 mL of the standard buffer in a 500-mL beaker and 100 mL of NaOH solution for the group. The third student should prepare the pH meter and make sure that the group has nine 100- to 250-mL beakers on hand for pH analysis.

Part II: Survey of Indicator Colors Over a pH Range

Each indicator should be tested in the tissue culture plates to determine where it undergoes a color change. This is done by taking the standard buffer and using it to prepare solutions at pH values near (± 0.1 pH unit) 4, 5, 6, 7, 8, 9, 10, and 11. Each student should have a tissue culture plate ready with one or two drops of indicator in each of eight wells.

As a group, take 100 mL of the standard buffer in a 250-mL or larger beaker. Check the pH value. This should be near pH = 4. Add NaOH in small amounts (about 0.50 mL). Mix well and record the pH value. When the pH is between 3.9 and 4.1, each student in the group should take some of this for the "pH = 4" well in his or her plate.

Continue adding base to the buffer. When the pH is within 0.10 pH unit of each of the target values, each student should again withdraw some buffer for a tissue culture plate well.

The eight solutions in the wells of the each tissue culture plate should show where the color changes dramatically over a range of two pH units. This will be your target for the spectrophotometric experiments.

Part III: The Spectrophotometric Titration

Prepare six clean, dry polystyrene cuvettes. *Use only water to clean these cuvettes. Acetone will dissolve the polystyrene!* Arrange them in a neat row. The cuvettes have two sides that are slightly cloudy and two sides that are clear. Label the cuvettes No. 1 through No. 6 on the cloudy side. Do the same with six small beakers or Erlenmeyer flasks.

Each student will now have to prepare six buffer solutions. One should be at least one pH unit below the transition range of the indicator (for example, at pH = 7 for phenolphthalein). The sixth should be one pH unit above the transition range of the indicator (for example, at pH = 12 for phenolphthalein). The pH of solutions 2 to 5 will span the center of the transition range (e.g., because phenolphthalein changes pH between pH = 8 and pH = 10, a suitable range would be pH = 8.5, 8.9, 9.2, and 9.5).

About 150 mL of the buffer should be taken by each student in a 400-mL beaker. Record the pH, then use the NaOH solution to adjust the pH to the lowest pH in the series. Record the pH, then use a volumetric pipet to transfer exactly 25 mL of this solution into the first small beaker or flask.

Next, adjust the pH of the remaining buffer to the lowest pH value for the transition range of the indicator. You should be within 0.10 pH unit of the target value. Record the pH, then use a volumetric pipet to transfer exactly 25 mL of this solution into the second small beaker or flask.

Continue to add NaOH to the buffer to reach a point within 0.10 pH unit of each of the values in the transition range. At each point, record the exact pH and use a volumetric pipet to obtain 25 mL for the third small beaker or flask. Conclude by adding NaOH to reach a pH at least 1.5 units above the highest value of the transition range.

Each student will now have six beakers or flasks with the buffer at the different pH values. Add *exactly* the same amount of indicator solution to each (0.50 mL is probably enough). Mix well, then put some of each into each of the cuvettes.

Each solution should be analyzed on the spectrophotometer. If a single-wavelength unit is used, set the instrument to the wavelength listed in Table C-2. If a recording spectrophotometer is used, then be certain to obtain the absorbance value at the wavelength in the table.

TABLE C-2

Indicator	Wavelength
Bromocresol purple	595 nm
Bromothymol blue	610 nm
Thymolphthalein	610 nm

REPORT

You should discuss in your report how all members of the group determined the pH range to use to study his or her indicator. Present a table listing the following data for each of the four intermediate samples you examined: pH of the sample, apparent color (that is, what you see), and absorbance at the reference wavelengths.

Prepare a plot of the absorbance values vs. pH for your spectra.

A second table should convert the values for the absorbances in cuvettes 2 through 5 into values for α and $\alpha/(1 - \alpha)$ and convert the pH's into [H^+]. The last column of this table should have a calculated K_a.

This experiment provides four independently determined values for K_a: one each from samples 2 through 5. Calculate an average and a standard deviation for K_a.

Experiment 3
APPLICATION LAB: DETERMINATION OF SERUM ALBUMIN

Pre-Laboratory Assignment   **Due Before Lab Begins**

NAME: _____

Complete these exercises after reading the experiment but before coming to the laboratory to do it.

1. During this laboratory, you will have to assess the "health" of a "patient" based on a sample of "blood." If you determine that this sample has an albumin concentration of 60 g L^{-1}, is the patient healthy or not? Explain your answer.

2. You mix 0.20 mL of a solution containing 6 g L^{-1} albumin with 2.80 mL of BCG reagent. Assuming volumes are additive, what is the concentration of albumin in the resulting solution?

3. Beer's Law is assumed to be valid in this experiment. Will Beer's Law apply if we choose to use concentrations in g L^{-1}? Explain your answer.

4. Since you are not working with human blood in this lab, is there any reason to worry about the safe handling of solutions in this lab? Explain your answer.

BACKGROUND

Now you are ready to apply spectrophotometry to the analysis of the serum protein albumin in blood. The method is based on the reaction of albumin with a dye, bromocresol green. At the pH of the experiment, bromocresol green is yellow, but when albumin is present, BCG forms a bright green complex with albumin. This reaction can be used in a crude assessment of albumin levels or in a precise automated method that uses the spectrophotometric techniques you have been learning.

Table C-3 indicates the expected range for albumin levels in the blood of healthy adult humans. Your method will permit the determination of albumin concentrations. In practice, other serum proteins, especially globulin, react with BCG. That will not be a problem in your experiments here, but in the clinic it is the reason why very precise albumin concentrations require more tedious and expensive methods. Nevertheless, initial screening for abnormal levels of albumin is effectively done with BCG, and then additional tests are done where required.

TABLE C-3

NORMAL RANGES FOR ALBUMIN AND IN THE BLOOD		
Protein	Normal Levels	Possible Values
Albumin	35–55 g L^{-1}	20–70 g L^{-1}

You and your group will develop two methods to determine albumin. One will be done with just a couple of drops of sample, similar to the finger-stick methods you may have seen in your own experience. To complement this, you will also find out how to determine albumin levels quantitatively. At the end of the lab period you will get albumin from a series of "patients," and use the approximate method to determine which require additional, quantitative, study.

Notes:

- As you know, you are not working with real human sera. All solutions are prepared from purified equine (horse) proteins. But, since this lab simulates the development of a method for human samples, you should practice good laboratory hygiene and do all manipulations using laboratory gloves. Gloves will be available for your use.

- Typical clinical albumin analysis is done at blood temperature, 37°C. We cannot provide these conditions for the spectrophotometers you will use, but we recommend that you experiment to see if warming the solution for 10 min in a 40°C water bath, followed by 1 min of cooling in room temperature water, makes any difference in the absorbance.

- The clinical analysis is done using very small syringes to measure serum amounts, typically 50 μL. Such syringes are not available for us, but small, 1.0-mL Mohr pipets are. To make it possible for you to measure solutions conveniently, all samples have been diluted by a factor of *40* relative to standard amounts. Thus, the stock solution we provide (2.50 g L^{-1}) will give a response equal to a patient with 100. g L^{-1}. If you want to measure a sample equivalent to 20 g L$^-$ in blood, you want to make a solution that has 20/40 = 0.50 g L^{-1} from the stock solution we provide.

- If you heat any solutions do not heat them above 40°C; otherwise you will cook the proteins!

 Caution: You will not be working with real blood samples. Your samples will be solutions of equine (horse) proteins. If you were using real blood samples, you would have to observe the necessary precautions for the safe handling of a blood sample. These include wearing latex gloves, disposal of the glassware used to contain the blood sample and the latex gloves in a biohazard bag after analysis is complete, and washing any spills with bleach.

In this lab, chemical-resistant gloves will be available. Their use is advised, since the BCG solution that you will use is a concentrated buffer solution.

Procedure

Part I: Formation of Groups

Groups of four will be responsible for testing the albumin–BCG reaction and each group member will prepare two solutions for the common calibration curve. Each member will have a unique set of unknowns to test.

Part II: Characterize the Albumin–BCG Reaction and Prepare a Calibration Curve

Each group will get an analysis kit containing the standard solutions of albumin (2.5 g L^{-1}) and bromocresol green. You will also find a solution of "standard" NaCl, which should be used for *all* protein dilutions and to fill out the volume required for the cuvettes.

It is important that the group members make observations together on albumin. Practice together adding a few drops of the albumin solution to about 1 mL of the BCG reagent. Watch carefully for differences in the response of the reagent. Check it with the spectrophotometer at 1 min and at 5 min after they are prepared. If you cannot record the entire UV–visible spectrum efficiently, then you may select the "best" wavelength by referring to the spectra in Figure C-7.

Once you have an idea of the response of the protein to the reagent, use standard solutions to design and test a quantitative method. First, make sure that you will always have enough BCG reagent by carrying out a microscale test. Add vary-

FIGURE C-7 UV–visible Spectra of Solutions of Albumin with BCG Reagent

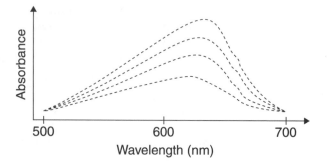

ing amounts of albumin to a *consistent* set of solutions of BCG and NaCl. Be sure that you never see any break in the color. Your preparation work will conclude with the group preparing eight solutions that cover the range of possible values for the concentration of serum albumin. Consult the list of possible ranges (not just healthy ranges) and prepare a 10.00-mL volumetric flask for each possible value of albumin. Use the absorbance of these solutions to get the calibration curve and to prepare the "reference sheet" for the finger stick analysis.

Part III: Preparation of a Finger-Stick Analysis

In clinical settings it is often important to get a variety of semiquantitative measurements in a very short time. This is done by wetting a piece of pretreated paper with a sample of blood. The change in the color of the paper indicates the approximate amount of the albumin (or other substance) that is present.

Simulate this using the microscale tissue culture plates in the laboratory. Prepare eight wells with the same amount of BCG and albumin solutions in each. Add the same amount (say, two or four drops) of one of the group's albumin standards to each and allow the color to develop. These make up your reference sheet for the next step.

Each individual will receive an assignment to test five different patients. The samples can be obtained from the eye dropper bottles. Transfer several drops of the sample for your assigned patients into a set of vials. Use microscale test to determine the concentration of albumin in each patient. Compare those to the reference sheet. Try to estimate the albumin to within 5 g L^{-1}.

Part IV: Spectrophotometric Determination of Serum Albumin

Two of the five samples you test will be for patients with either high or low albumin levels. These must be tested further before medical treatment is begun! Return to the "patient" and get some more sample, but not more than 1 mL. Use this, along with the calibration curve and the spectrophotometer, to determine the concentration of albumin to within 2 g L^{-1}.

CALCULATIONS AND REPORT

During the Laboratory

Carry out a quick analysis of the unknown you are studying. Determine the presence of low, normal, or high concentrations of albumin in each of the five "patients," and then the concentration in grams per liter of albumin. Turn in the "stat" report form to your instructor to indicate whether the patient is in danger because of an abnormal level of albumin.

After the Laboratory

Your report should start with a one-paragraph standard procedure that you think can be used to analyze samples in the future. Next, include a section discussing any restrictions on the method: what ranges of albumin will *not* work with this method, and any information about how to handle the sample. Finally, present one paragraph indicating why this method may be very helpful in rapid determination of albumin and another paragraph indicating why it might *not* be helpful.

EXPERIMENT GROUP D

ACID–BASE TITRATION AND THE GLOBAL CARBON CYCLE: PREDICT EFFECTS OF RISING LEVELS OF CO_2

Collecting plant material is a first step in many ecological investigations.
D. Thomas/Visuals Unlimited.

PURPOSE This is a three-week set of experiments to give you experience in acid–base titration and its use in analysis of mixtures. During the first week you will work individually on acid–base titrations, including the standardization of solutions. The second week you will become familiar with acid–base titrations and back titrations as you analyze different antacids. In the third week you will use these techniques in the analysis of the amount of CO_2 produced by decomposing leaves.

The scenario for this module is taken from the actual methods employed by field ecologists when they need to determine how rapidly biomass on the forest floor—especially leaves—is decomposing. This is an important factor in the balance of carbon dioxide in the atmosphere, and this in turn may be important in climate control.

SCHEDULE OF THE LABS

EXPERIMENT 1: Skill Building Lab: "Acid–Base Titration"

1. Determine the concentration of a solution of NaOH using KHP as a primary standard (individual work).
2. Determine the concentration of HCl and H_2SO_4 using NaOH as a secondary standard (individual work).

EXPERIMENT 2: Foundation Lab: "Analysis of an Antacid"

1. Analyze different antacids by neutralization with an excess of a solution of HCl and titration of the unreacted acid with NaOH (group work).
2. Determine the neutralizing capacity of an antacid as moles of hydrogen ion (n_{H+}) per gram of tablet. This is the ratio n_{H+}/g (individual work).
3. Compare the effectiveness of different antacids (group work).
4. Set up two leaf compost chambers and a control jar for the third week of the experiment (individual work).

EXPERIMENT 3: Application Lab: "Tree Leaves and the Global Carbon Cycle"

1. Test the solubility of metal cations to determine which is best for the selective precipitation of carbonate in the presence of hydroxide (group work).

2. Test the method for the determination of NaOH in a $NaOH/Na_2CO_3$ mixture. Add excess metal cation solution to the mixture then titrate (group work).

3. Use the procedure just tested to determine the number of moles of unreacted NaOH in the experimental setups (leaf compost jars and control jar—individual work).

4. Calculate the number of grams of CO_2 per gram of dry leaf per day under varied conditions (individual work).

SCENARIO You have a summer internship! A biologist wants to use your knowledge of chemistry to help predict the effects of rising atmospheric CO_2. She tells you that the CO_2 concentration in the atmosphere is rising because of the burning of petroleum, coal, and wood. She says that predicting the effects of an atmosphere with a high CO_2 concentration on society is important. It is possible to imagine, she says, negative effects such as climate warming leading to sea-level rise and coastal flooding. But, she continues, it is also possible to imagine positive effects such as improved plant growth and more land in the north available for agriculture.

The biologist shows you the equation

$$C_n(H_2O)_m + nO_2 \rightarrow nCO_2 + mH_2O + \text{energy}$$

FIGURE D-1 THE GLOBAL CARBON CYCLE; FIGURES IN PARENTHESES REFER TO BILLIONS OF TONS (AFTER J. E. FERGUSON, *INORGANIC CHEMISTRY AND THE EARTH*; PERGAMON PRESS: LONDON, 1982)

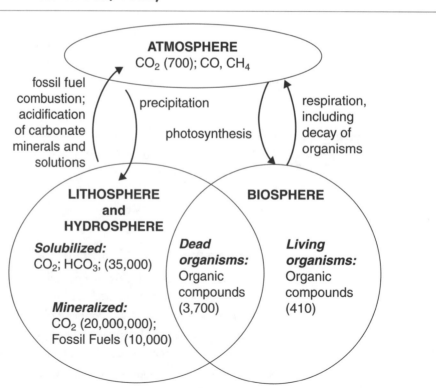

She grows plants in chambers containing elevated partial pressures of CO_2 in order to study the right-to-left reaction. She suggests that you study the left-to-right reaction, which gives the amount of CO_2 returned to the atmosphere by decomposing leaves. She says there is great interest in whether some of the carbon fixed by plants fails to decompose and is thus kept out of the atmosphere. And with that, she goes back to her own work.

But what exactly should you do? To decide, you spend some time reading about the effect that fossil fuel use has on the carbon cycle in the library (see Figure D-1). You learn two things. First, almost 90% of the carbon returned to the atmosphere is from the respiration of decomposers (bacteria and fungi) breaking down dead leaves. Second, the burning fossil fuels release other substances in addition to CO_2. One of these other substances is nitrate, which like CO_2 is a plant nutrient.

You decide to do two experiments that will answer the biologist's question and a question that formed in your mind as you read in the library. First, the biologist asked whether some of the carbon fixed by plants fails to decompose. You don't have time to study wood decomposition, and you can't visit peat bogs where carbon probably is accumulating. So you decide to use leaves and ask whether they could decompose in a year under favorable conditions. You reason that if the number of days it would take leaves to decompose under favorable lab conditions is more than a year, than carbon will accumulate and there will be less in the atmosphere.

Your library reading made you curious about a second question that your biologist supervisor didn't say anything about. The nitrate that is released from burning fossil fuels is an important plant nutrient. Could nitrate increase the decomposition rate? If so, that would be a reason to expect rapid decomposition of leaves and little accumulation of carbon in dead plant material.

Experiment 1
SKILL BUILDING LAB: ACID–BASE TITRATION

pg 7 D for help

BACKGROUND

In this experiment, you will titrate with *both* a primary standard and a secondary standard. The primary standard will be used to determine the concentration of the secondary standard. The secondary standard will be used to determine the concentration of a third chemical.

In Part I of the experiment you will prepare and use solutions of the primary standard KHP (potassium hydrogen phthalate, $C_8H_5O_4K$) (see Figure D-2) to determine the concentration of an aqueous solution of NaOH. KHP is a monoprotic acid. The acidic hydrogen is attached to an oxygen in the molecular structure of KHP drawn here. The KHP has been dried in an oven to drive off any water that might contaminate it. In Part II you will use the NaOH you studied in Part I as a secondary standard in a reaction with HCl. Finally, in Part III you will again use your secondary standard in a determination of the concentration of a solution of sulfuric acid, H_2SO_4.

The indicator for use in this experiment has been chosen for its sharp end point when only a fraction of a drop of excess NaOH has been added to a solution of an acid. The indicator is *phenolphthalein*. It turns from colorless to a faint pink at the end point and goes on to a dark pink or purple color when the end point is overshot.

 Caution: You will be working with acids and bases in this experiment. While they are dilute solutions, they are corrosive. Wash all small spills (drops) with water. Notify your instructor in the event of a large spill (several milliliters). Always wear appropriate safety goggles and wash your hands thoroughly before leaving the lab.

PROCEDURE

NOTE: All work in this experiment is to be done on an individual basis.

Part I: Standardization of a Solution of Sodium Hydroxide Using a Primary Standard.

Clean and rinse well (with deionized water) four 125-mL or 250-mL Erlenmeyer flasks. These do not have to be completely dry.

Measure between 0.20 and 0.25 g of potassium hydrogen phthalate into each Er-

FIGURE D-2 MOLECULAR STRUCTURE OF POTASSIUM HYDROGEN PHTHALATE, KHP

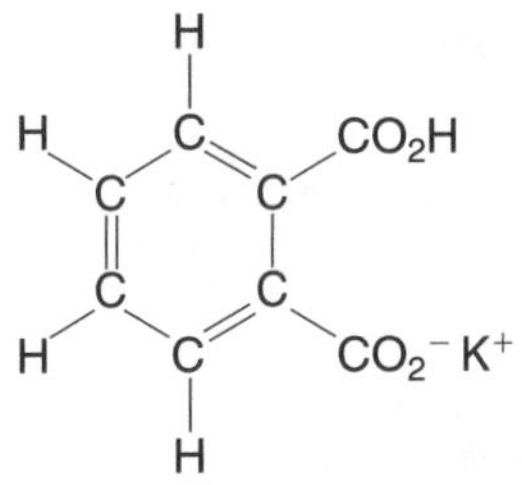

lenmeyer flask. Record the *exact* weight added to each flask. Add about 25 mL of water to each sample. Set the samples aside.

Find out from your instructor which unknown you will use. Obtain about 350 mL of the stock NaOH solution in a clean, dry beaker. Make a note of its unknown code if more than one unknown is used for your laboratory.

Prepare the buret as discussed in the Introduction. Remember to make sure the buret is clean, to rinse it with deionized water, and to prerinse with the NaOH solution you will use.

By this time, the KHP samples should all be dissolved. If not, swirl their flasks gently. When you are sure they are fully dissolved, add two drops of phenolphthalein indicator to each. The solution should be colorless.

This first titration gives you a rough idea of the amount of NaOH you need to neutralize the KHP. It should be done quickly by adding approximately 0.50-mL portions of the NaOH solution, then swirling the solution. The approach to the end point is suggested by the temporary appearance of a pink color that fades when the solution is swirled for up to 10 s. A pink color that persists for at least 30 s signals the actual end point.

Repeat the titration on the second, third, and fourth samples. For these you can add a significant portion (70–80%) of the NaOH solution quickly and then, near the end, add the NaOH *one drop at a time* until you see the end point.

> **NOTE:** For the most accurate results, try to add *fractions* of a drop. Open the buret stopcock to allow a small amount of solution to form a bead on the tip. Then wash that down into the titration flask with a gentle stream of deionized water.

Part II: Determination of the Concentration of a Solution of Hydrochloric Acid Using the NaOH Secondary Standard

A calculation of the concentration of the NaOH gives you a secondary standard to use in determining the concentration of the hydrochloric acid solution. You do not need to do this calculation during the lab.

Your instructor will tell you which of the HCl solutions to use. Use a clean dry beaker to obtain about 125 mL of your assigned solution from the stock bottle.

Clean and rinse well (with deionized water) four 125-mL or 250-mL Erlenmeyer flasks.

Obtain a volumetric pipet and record its volume. Use this to transfer the given volume of HCl solution into each of four flasks. Add two drops of phenolphthalein indicator to each.

Titrate the HCl solution with the NaOH from the buret. Again, expect that you will overshoot the end point on the first run. Make an effort to be much more accurate with the second, third, and fourth trials.

Part III: Determination of the Concentration of a Solution of Sulfuric Acid Using the NaOH Secondary Standard

Use a clean dry beaker to obtain about 125 mL of your assigned sulfuric acid solution from the stock bottle.

> **HINT:** The concentration of sulfuric acid you use in this part will be very close (within 2%) to half of the concentration of HCl you used in Part II.

Clean and rinse well (with deionized water) four 125-mL or 250-mL Erlenmeyer flasks.

Obtain a volumetric pipet and record its volume. Use this to transfer the given volume of sulfuric acid solution into each of four flasks. Add two drops of phenolphthalein indicator to each.

Titrate the sulfuric acid solutions with the NaOH from the buret. Again, expect that you will overshoot the end point on the first run. Make an effort to be much more accurate on the second, third, and fourth trials.

REPORT

Present your results in a table. You must give concentrations for your assigned solutions of NaOH, HCl, and H_2SO_4. Indicate the amount of primary standard used in each of the four trials. Show sample calculations used to determine the concentration of

a. The secondary standard NaOH.
b. The hydrochloric acid solution.
c. The sulfuric acid solution.

Report your final results as an average of the second, third, and fourth titrations in each part. Briefly (three sentences) summarize your observations (color changes in particular—permanent or temporary) for each of the three titrations in each part. For each solution analyzed, account for any differences between the calculated molarity for each trial.

Experiment 2
FOUNDATION LAB: ANALYSIS OF AN ANTACID

Pre-Laboratory Assignment   **Due Before Lab Begins**

NAME: _____

Complete these exercises after reading the experiment but before coming to the laboratory to do it.

1. Why do you rinse the buret with the solution you will be dispensing from the buret?

2. What do think will happen on a molecular level when you add the antacid to the HCl solution? Why?

3. What purpose does the indicator serve?

4. After adding the antacid to the HCl in this experiment, do you expect the solution to be acidic, basic, or neutral? Explain.

5. What precautions should you take while heating the antacid–acid mixture? Why?

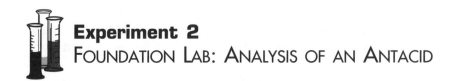

Experiment 2
FOUNDATION LAB: ANALYSIS OF AN ANTACID

BACKGROUND

Before you can analyze CO_2 from dead leaves, you must first develop and test a method of analysis. You will test this method on an antacid tablet, which will respond similarly to your method as will your dead leaves.

A popular ad shown several years ago pictured a person in obvious discomfort saying "I can't believe I ate the whole thing." If you have ever found yourself in a similar situation, do you respond as the person in the ad did by popping an antacid? What does an antacid do to relieve your discomfort? The pH in your stomach is normally 2 to 3. When your stomach pH level drops, you may take an antacid to neutralize the excess acid and return the pH level to normal (not all the way to a pH of 7, you don't want to neutralize *all* of the acid).

If you use an antacid, how did you decide which brand to buy? In this experiment you will compare different brands of antacids. Are all antacids equally effective in neutralizing acid? This is what you will determine in this experiment.

To determine how much acid an antacid tablet can neutralize, you will dissolve the antacid in *excess* acid and then titrate the unreacted acid with a standard NaOH solution. You will know how much acid you started with and how much of it reacted with the NaOH, so you will be able to determine how much reacted with the antacid.

 Caution: You will be working with acids and bases in this experiment. These solutions are corrosive. Wash all small spills (drops) with excess water. Notify your instructor in the event of a large spill (several milliliters). Always wear appropriate safety goggles and wash your hands thoroughly before leaving the lab.

PROCEDURE

Part I: Assign Tasks to Group Members

We recommend working with no more than three people per group. Review the Introduction about the formation of groups. Each group should analyze a different antacid, and each member should perform two analyses so that the group has at least six results on the same antacid. Then each group will meet as a group to discuss results.

Part II: Acid–Base Titration of Antacid (Group Work)

Obtain at least two burets for the group, one for use with HCl and the others for NaOH. Prepare the burets as discussed in the Introduction. Remember to make sure the burets are clean, to rinse them with deioinized water, and to prerinse with the solution they will each contain. Label the burets so you know which are for the NaOH and which is for the HCl.

Determine the mass of each of your antacid tablets. When you obtain the HCl and NaOH, record their exact molarity as they appear on the stock bottles. This will vary from day to day.

Dispense no more than 25 mL of HCl solution into each of two clean Erlenmeyer flasks. Record the exact volume of HCl used in each. Then add one of the antacid tablets to each flask and heat gently for 1 or 2 min to help to dissolve the tablets. Bring each solution to a boil to dispel any undissolved CO_2. Cool. Some solid materials, used to bind the antacid tablet together, may result.

Add two or three drops of an indicator to each of your antacid mixtures. See your instructor if they are not the color you expect keeping in mind that you want an excess of acid present in each mixture.

Titrate each antacid mixture with the NaOH solution until you have reached the end point of the indicator used.

Calculations

Calculate the moles of hydrogen ion neutralized by the antacid per gram of antacid tablet (n_{H^+}/g). Determine an average value for the antacid you analyzed.

Write the balanced chemical equation for the reaction between HCl and the active ingredient in your antacid tablet, then calculate the mass of the active ingredient per tablet in your antacid, based upon your data.

RESULTS AND CONCLUSIONS

Individual Report

Compare the precision of your own trials with each other and with the other members of your group. Account for differences.

How does the calculated mass of the active ingredient compare with the information on the label? Account for differences.

This laboratory uses a *back titration*. How is the method here different from the direct addition of HCl to the antacid? Why do you think a back titration is used here?

Group Report

Discuss the following and prepare a final group report.

1. Determine the group average n_{H^+}/g of antacid. If the group decides to eliminate any values from the group average, justify the decision to do so.
2. Calculate the average mass of the antacid tablets.
3. Calculate the cost of the antacid per gram in cents per gram.
4. Report the ingredients in your antacid, the average n_{H^+}/g of a tablet, and the cost per gram to the entire class.
5. After all the class data are tabulated, discuss which antacid you would buy and why. Are there considerations besides cost and neutralizing ability to consider?
6. Record the results of all group discussions and hand in a final group report.

PREPARATION FOR WEEK 3 EXPERIMENT

Set up your two leaf compost jars and a control jar and leave all three jars in the same location until next week.

1. Obtain two sets of three or four leaves each (use the same type of leaf in both jars, i.e., maple, oak, etc.), three airtight jars, three small beakers, and a 20-mL pipet.
2. Weigh each set of the dry leaves separately.
3. Using the pipet, measure 20.0 mL of the approximately 1.00 M NaOH that you used in the antacid experiment into each of the three small beakers. Record the exact molarity of NaOH solution.

4. Place one of the small beakers containing NaOH into one of the jars and seal the jar.

5. Place one set of leaves into one of the other jars. Spray some water on the leaves, then place the second small beaker with NaOH on top of the leaves. Be careful to avoid direct contact between the leaves and the NaOH solution. Seal the jar.

6. Place the second set of leaves into the remaining jar. Spray some water containing dissolved nitrate on the leaves, then place the last small beaker with NaOH on top of the leaves. Be careful to avoid direct contact between the leaves and the NaOH solution. Seal the jar.

7. Place all three jars in the location specified by your instructor.

Experiment 3
APPLICATION LAB: TREE LEAVES AND THE GLOBAL CARBON CYCLE

Pre-Laboratory Assignment   **Due Before Lab Begins**

NAME: _____

Complete these exercises after reading the experiment but before coming to the laboratory to do it.

1. What is the purpose of a control in an experiment? Which sealed jar serves as the control in this experiment? How do you know?

2. Is your mixture in Part III a solution of acids or bases? What color is the phenolphthalein in this mixture?

3. Write balanced equations for the following:

   a. Hydrochloric acid and sodium hydroxide.

   b. Hydrochloric acid and sodium carbonate.

   c. Sodium hydroxide and carbon dioxide.

4. Explain what you would do if you inadvertently left the stopcock open while filling the buret with the HCl solution and it spilled out over your bench top.

Experiment 3

APPLICATION LAB: TREE LEAVES AND THE GLOBAL CARBON CYCLE*

BACKGROUND

Before you analyze your decomposing leaves, you need to determine how you are going to complete the analysis. Your goal in this experiment is to determine the average mass of CO_2 produced per gram of leaf per day. Each sealed jar contains a beaker of sodium hydroxide. Sodium hydroxide solutions absorb CO_2 to produce sodium carbonate and water. CO_2 is present in the atmosphere at all times. The jar without the leaves will reveal the presence of CO_2 upon analysis, and this will serve as a control to determine how much of the CO_2 in the leaf jar is really due to leaf decomposition.

After time, the solution of sodium hydroxide in the jars will contain sodium carbonate along with the unreacted NaOH, as a result of absorption of CO_2. Assuming you started out with an excess of sodium hydroxide, you can calculate how much CO_2 reacted by determining how much sodium hydroxide remains unreacted. This will be accomplished by titrating the sodium hydroxide with a solution of hydrochloric acid *after* you remove the carbonate from solution. Of course, HCl will eventually react with the solid carbonate: your technique must allow you to find the end point for the reaction of HCl with NaOH before the HCl begins to react with the carbonate.

Part II of this experiment will enable you to determine which reagent is best for the selective precipitation of the carbonate ions in the presence of hydroxide ions in solution. Part III gives you practice with accurate titration of NaOH in the presence of precipitated carbonate ions. Finally, in Part IV, you will analyze your experiment with decomposing leaves.

Caution: You will again be working with acids and bases in this experiment. These solutions are corrosive. Wash all small spills (drops) with excess water. Notify your instructor in the event of a large spill (several milliliters). Always wear appropriate safety goggles and wash your hands thoroughly before leaving the lab.

One of the test solutions in Part II may contain a barium compound (it will be labeled if you have it). Many barium compounds are *poisonous*. Handle with care. Dispose of all barium-containing solutions in the designated waste receptacles.

PROCEDURE

Part I: Formation of Groups

Groups of two or three may be used in this experiment. Do *not* begin work on your leaf experiments or your control experiment until you are certain of the best procedure to determine the amount of carbon dioxide. You will only have one chance with each.

Part II: Selective Precipitation of Carbonate Ions (Group Work)

---

* The method used to analyze for CO_2 from the decay of leaves is described in Anderson, J. P., "Soil Respiration," pp 831–872 in A. L. Page, editor, *Methods of Soil Analysis*, Part 2, 2nd Ed. Soil Science of America and American Society of Agronomy: Madison, WI, 1982.

Determine which of the available test solutions reacts to form an insoluble precipitate with sodium carbonate but not sodium hydroxide. Since you will be using a spot plate for your tests, you only need to use about 10 to 20 drops of each of the solutions per test. Record all observations.

Results and Conclusions for Part II

Based upon your observations, decide which reagent will selectively precipitate the carbonate ions in the presence of the hydroxide ions (thus leaving the hydroxide ions in solution). Write the net ionic equation for the reaction between the carbonate ion and the reagent you select.

Part III: Titration of NaOH/Na$_2$CO$_3$ Mixtures (Group Work)

Prepare a buret for use with HCl.

Obtain a solution containing a mixture of sodium hydroxide and sodium carbonate solutions in unknown amounts. These solution mixtures were made by mixing 1.0 M solutions. Transfer a known quantity of the solution mixture to a 125-mL Erlenmeyer flask. Add two or three drops of phenolphthalein indicator.

Titrate the mixture with the HCl until the indicator color just disappears. The phenolphthalein color fades if left alone because it oxidizes in air. To be sure you have reached the end point, add another drop of phenolphthalein. Record the volume of HCl needed to reach the end point.

Next, transfer another sample of the sodium hydroxide–sodium carbonate mixture to a clean, dry Erlenmeyer flask. Be sure to measure the same volume as used in the first titration. This time, before titrating with HCl, precipitate out the carboate by adding an excess of the metal cation solution you selected in Part II to precipitate the carbonate selectively. To determine how much is "excess," assume the total volume of solution used is only sodium carbonate and calculate the volume of the cation solution needed to completely react with this volume of 1.0 M sodium carbonate. Record your observations.

Add two or three drops of phenolphthalein indicator to your mixture and again titrate with HCl until the color disappears. Verify you have reached the end point by adding another drop of phenolphthalein.

Calculations for Part III

Calculate the percentage, by volume, of NaOH solution in your unknown mixture.

Results and Conclusions for Part III

Compare the volume of HCl used in both titrations. Explain any differences. If you are not within 5% of the expected value for the %NaOH, repeat your analysis on a different sample of the same unknown. See your instructor for the expected %NaOH of your unknown.

Part IV: Analysis of the Decomposing Leaves Systems

As the bacteria and fungi on the leaves respired, they produced CO_2. The CO_2 was absorbed into the sodium hydroxide solution, producing sodium carbonate and water. Based upon your work in Parts II and III of this experiment, determine how much sodium hydroxide remains unreacted. Remember to treat the control in the same way you treat the experimental setups.

CALCULATIONS

Write equations for each of the reactions which occurred in this section of the experiment. You should have three different equations. Calculate the mass of CO_2 produced per day per gram of dry leaf in each of the two jars. Remember, CO_2 is present in the jar even without the decomposing leaves. Your job is to determine how much *more* CO_2 is produced as a result of the leaf decomposition.

RESULTS AND CONCLUSIONS

Post your group's results for the entire class. Include the type of leaf used and the mass of CO_2 per gram of leaf per day for both plain water and the nitrate-enriched water. Discuss the results in class. Include the class results in your report. Discuss how different conditions may have influenced the results.

The Carbon Cycle

To understand more about the atmospheric levels of carbon dioxide, visit the following site on the World Wide Web. Be careful to get the case of all the letters correct.

http://ingrid.ldgo.columbia.edu/SOURCES/KEELING/

Click on "Views" to see the pattern of atmospheric CO_2 levels in the northern hemisphere. Describe the pattern observed on the graph. Return to the main page at the Web site and click on "Tables." Select "CO_2" as the variable you would like to make a table from. Next click on "columnar table." This table will enable you to compare the net CO_2 uptake with net return to the atmosphere via respiration. Each member in your group should select two different years. Prepare a table containing the atmospheric CO_2 level (in ppm) for each month of your years (record the values directly from the Web site). Share your data with group members.

QUESTIONS TO ANSWER IN YOUR REPORT

Include your responses to these questions in your report and then write a summary report to your biologist supervisor explaining your results and your responses to the original questions posed in the scenario.

a. Using your data and the fact that plants are roughly 44.5% carbon, calculate how many days it would take for all of the carbon in your leaves to be respired if microbes continue to metabolize at the rate you measured. Now think back to a just-fallen leaf outside. Considering how many days you estimate it would take for your leaves to disappear, do you think there are enough days that are moist and warm enough to allow complete metabolism of leaves between fall and spring, say between November 1 and May 15? Explain.

b. Using the data in your table from the Web site, describe the variations in the concentration of CO_2 in the atmosphere throughout the year. Circle the highest and lowest values each year.

c. What two biological processes are responsible for seasonal increases and decreases in atmospheric CO_2 concentration? Atmospheric CO_2 concentrations are a balance between these two processes.

d. Atmospheric CO_2 is expressed in ppm by volume on the web site. Assuming the atmosphere contains 1.8×10^{20} mol of air, calculate the *net mass* of CO_2 that was added and withdrawn from the atmosphere in a given year.

e. Returning to the data at the Web site, is the difference between the annual high and low atmospheric CO_2 concentrations the same from year to year? (Look at the whole table.) Explain.

f. Is there any evidence that atmospheric carbon is accumulating on land? Use parts (a) and (e) as evidence.

g. Gasoline and other fossil fuels are primarily hydrocarbons. Therefore, when fossil fuels are burned, CO_2 is produced. Assume that gasoline is composed entirely of octane, C_8H_{18}, and that the burning of gasoline (as in your car engine) can be represented by an equation for the complete combustion of a hydrocarbon. Determine the mass of CO_2 produced by your car in a year. Be sure to record any other assumptions you make in determining your answer. The density of octane is 0.692 g/mL. If you do not operate a motor vehicle, assume you get 19 miles per gallon and drive 11,000 miles per year.

EXPERIMENT GROUP E

BUFFERS AND LIFE: SAVE A CARDIAC ARREST PATIENT

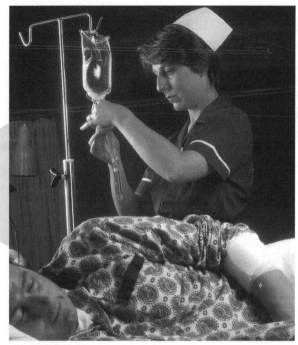

Properly adjusting a patient's fluids can be critical to survival.
James Holmes/Science Photo Library/Photo Researchers.

PURPOSE The chemistry of water includes the very important property of acidity and basicity. A solution's acid–base properties can dominate the way that other substances exist and react in water. One measure of the strength of an acid or a base is the amount of hydrogen ion in the solution. This can be followed as a concentration, commonly written $[H^+]$, with units of moles per liter.

The wide range of possible values for $[H^+]$ means it is often easier to follow a derived function, the pH, of the solution. The pH is a function of the concentration of hydrogen ion, H^+, in the solution, where $pH = -\log[H^+]$. At low pH the concentration of hydrogen ion is high and many substances acquire a hydrogen ion. At high pH many substances will lose a hydrogen ion. This can radically alter the properties of the substance, something that is very obvious with the color of acid–base indicators. It is no surprise, then, that the pH value of a biological system is critical to its operation.

The importance of pH means also that it is important that we know how to stabilize pH. This is done very conveniently with a **buffer**. An acid–base buffer is a mixture that can act to keep the pH of a solution at nearly a constant value. Some buffers can be prepared at very high or very low pH values using strong acids or bases. But the most common kind of buffers, the subject of this experiment group, stabilizes the pH at intermediate values—between $pH = 2$ and $pH = 11$. These buffers form when both a weak acid and its conjugate weak base are present in solution in significant concentrations.

In this group of labs you will investigate how buffers work, then apply your understanding to the design of a solution that has certain pH properties. Finally, you will act to correct the pH of a buffer that emulates the blood of a victim of cardiac arrest, since cardiac arrest can cause the pH of the body to change by a dangerous amount.

SCHEDULE OF THE LABS

EXPERIMENT 1: Skill Building Lab: "Stabilization of pH with Buffers"

1. With a pH meter, follow the effect of added strong acid and strong base on the pH of pure water and on a solution of a buffer (individual work).
2. Compare the different ways to determine the "strength" of an acid or a base (group work).

EXPERIMENT 2: Foundation Lab: "Design of a Buffer"

1. Using a mixture of a weak acid and its conjugate weak base, make a buffer that has a particular pH and a particular strength (individual work).
2. Test buffers for pH changes when strong acid and strong base are added (group work).

EXPERIMENT 3: Application Lab: "Cardiac Arrest! The Restoration of Buffers"

1. Assess the buffering ability of solutions for healthy adults (group work).
2. Characterize the changes in the pH when a weak base is added to a buffer at unhealthy pH values (group work).
3. Given a solution buffered at an unhealthy pH value, add an appropriate amount of the conjugate base to restore the buffer (individual work).

SCENARIO

In a hospital emergency room, a patient has just been wheeled in following a cardiac arrest in his home. The paramedics were able to restart his heart after a couple of minutes, and prior to that CPR was administered so that enough oxygen flowed to his tissues to keep them alive. But his face has the ashen color of someone close to death, and his skin is cold. Though his heart is beating again, he is in danger of death from shock. Immediately, one emergency room nurse starts to prepare an intravenous tube, and another gets a bag of fluid from the supply cabinet. The doctor says, "Yes, let's start him on bicarb" and the nurses begin to administer the fluid, one that helps to restore the victim's blood to its normal state and staves off further damage.

The fluid the ER team has just used to start the stabilization of the patient is the same one that you might use to quiet an upset stomach, or to mildly cleanse a soiled surface. It is sodium hydrogen carbonate (or bicarbonate of soda, known also as "bicarb" or "baking soda."). The formula of sodium hydrogen carbonate, $NaHCO_3$, is quite simple, but its role in the body is profound. The hydrogen carbonate ion, HCO_3^-, works with its conjugate acid carbonic acid, H_2CO_3, to form a stable pH buffer in the blood. This keeps the blood at a constant pH, something that is essential for life.

The pH of healthy adults is usually between 7.35 and 7.45. This is maintained by a 1:20 ratio* of $H_2CO_3:HCO_3^-$. Very small changes in the amount of either substance can radically alter the pH and result in sickness and death. Some of these changes are associated with problems in breathing, such as pneumonia and hyperventilation. In this experiment group, however, we will be more concerned with a patient with metabolic problems.

---

* This ratio depends on normal body pressures and temperatures. While we work at lower temperatures in the laboratory where the ratio is different, the 1:20 ratio is still a helpful guideline.

If the ratio of carbonic acid to hydrogen carbonate becomes too low, then the blood pH will rise and the body will become alkalotic. This can happen, for example, when vomiting occurs and the body uses up carbonic acid to reestablish the stomach acid.

The ratio of carbonic acid to hydrogen carbonate can also be too high, which happens when excess acid is present in the body. The blood pH drops and the body becomes acidotic. Acidosis occurs when, for example, the body is unable to remove carbon dioxide, because the blood stops circulating in cardiac arrest.

When the body starts to become acidotic or alkalotic, there is often enough time for a corrective measure. Most cases of blood pH imbalance are treated by allowing, or by inducing, the body to make the necessary adjustments. This process is known as compensation by the blood. Compensation for pH problems does not correct the ultimate source of the imbalance. But it can prevent other problems from arising.

Certain events cause such dramatic and threatening pH changes that we cannot wait for natural compensation. When a cardiac arrest occurs, the blood stops moving through the body. Immediately, carbonic acid starts to build up, lowering the pH. In addition, normal metabolism halts and lactic acid, the product of partial metabolism of carbohydrates, starts to build up. This, too, causes the 1:20 ratio to rise. Even after the heart is started, the change in the blood's buffer may be so severe that it places the patient in grave danger. Administration of hydrogen carbonate is sometimes the only way to ensure that further, perhaps fatal, damage does not occur. Consequently, an understanding of blood pH and buffers is essential to proper methods in the emergency room.

In these experiments you will progress from studying how buffers work to the point where you are ready to simulate the time-sensitive nature of an emergency room procedure. The material you learn in week 1 will be essential if you are to save your "patient" in week 3.

Experiment 1
SKILL BUILDING LAB: STABILIZATION OF pH WITH BUFFERS

BACKGROUND

Many different technical systems require the control of conditions so that radical changes do not occur when the system is stressed in some way. Here is just a short list, from very different areas of activity:

Activity	Possible Stress	Control Mechanism
Temperature of a baking oven	Room-temperature batter placed in the oven.	Large stone "baking tiles" placed in the oven
Suspension bridge	Traffic, some heavy, some light, on the roadway	Suspension cables can stretch a little
Government	Changes in laws	Constitutional guarantees of rights
International commerce	Currency fluctuations	Central banks choose to buy and sell reserves
Body temperature	Hot and cold days	Perspiration, increased metabolism

In all of these activities, something can smoothly act to correct the stress, whether it is in one direction or another. The correction may not be complete, but it will lessen, or *buffer* the system against the stress. However, in each and every case a large enough stress can overwhelm the control mechanism. In other words, the buffer is not an absolute barrier to change; it is a way of managing stresses of a certain scale.

Chemists apply the concept of buffering control to solutions with certain acid–base properties. Mixtures that are able to resist changes in the concentration of hydrogen ion are called acid–base **buffers.** Buffers keep the hydrogen ion concentration, and hence the solution's pH, from varying much. In some cases, our lives may depend on it.

How do we measure a buffer's capacity to resist change, and in what range does a buffer do the "best" job of stabilizing the pH? This skill-building lab will let you observe buffers in action, and lead to an understanding of how they are characterized, how they work, and how they fail.

Aqueous Acids and Bases

The acid and base properties of water come from the presence of two special ions, hydrogen ion H^+ (this is sometimes written as a complex with water called hydronium, H_3O^+) and hydroxide ion, OH^-. Ordinary water contains a small amount of both hydrogen and hydroxide ions. The reason for this is that water ionizes to a limited extent according to the equation

$$H_2O \ (l) \rightleftarrows H^+ \ (aq) + OH^- \ (aq)$$

or

$$2 \ H_2O \ (l) \rightleftarrows H_3O^+ \ (aq) + OH^- \ (aq)$$

This is an equilibrium reaction. That means that the reaction only proceeds so far. This is expressed mathematically by an equilibrium expression and an equilibrium constant:

$$K_w = [H^+][OH^-] \qquad\qquad K_w = 1.01 \times 10^{-14} \ (25°C)$$
equilibrium expression $\qquad\qquad$ *equilibrium constant*

If we know the concentration of either hydrogen ion or hydroxide ion, then we can calculate the concentration of the other using this expression.

In pure water, the concentrations of hydrogen ion and hydroxide ion are the same, but other chemical substances can affect these. To understand this we use the Arrhenius definition to define an acid as any substance that produces hydrogen ions in water. Solutions of acids will have more hydrogen ions than pure water. A base is any substance that produces hydroxide ions in water. Basic solutions have more hydroxide ions (and less hydrogen ions) than are normally present in pure water.

There are many substances that undergo reaction in water to form hydrogen ion. The process by which an acid reacts with water to form hydrogen ions is called *ionization* of the acid. Some acids, called **strong acids,** undergo complete ionization when they dissolve. This means that, for every molecule of the acid that enters the solution, one hydrogen ion is formed immediately. These acids are called strong because they give all the hydrogen ions they can as soon as they dissolve. When one mole of a strong acid is added to water, then one mole of hydrogen ions forms. Note that some strong acids release a second hydrogen ion in a subsequent step that does not occur when the strong acid alone is in water.

Some acids do not react completely to give hydrogen ion when they are placed in water. These are called **weak acids.** Some of the weak acid remains un-ionized, and some of the weak acid ionizes to form on less than one mole of hydrogen ion per mole acid.

Strong acid: $\quad$ HA $(aq) \rightarrow H^+ \ (aq) + A^- \ (aq)$

Weak acid: $\quad$ HA $(aq) \rightleftarrows H^+ \ (aq) + A^- \ (aq)$

Some strong bases react completely in water to give hydroxide ions. For the purposes of this lab, there is only one group of strong bases to consider: hydroxide salts. Many compounds, especially the alkali (group I or 1) metal hydroxides and the alkaline earth (group II or 2) hydroxides, are strong bases. Strong bases give a stoichiometric amount of hydroxide ions when they dissolve in water. This means that *every available hydroxide* in the salt becomes hydroxide ion in the solution, whether the salt contains one, two, or three OH^- ions.

$$NaOH \ (s) \rightarrow Na^+ \ (aq) + OH^- \ (aq)$$

$$Ba(OH)_2 \ (s) \rightarrow Ba^{2+} \ (aq) + 2 \ OH^- \ (aq)$$

There are also weak bases. These react with water to give hydroxide ion, but only partially:

Weak base: $\quad$ A$^-$ $(aq) + H_2O \ (l) \rightleftarrows OH^- \ (aq) + HA \ (aq)$

Note that this reaction is *not* the reverse of the reaction of a weak acid with water.

The pH Function

Chemists have found it much easier to express the acid content of a solution by using a logarithmic scale called the pH scale. For a base ten number system such as the one we use, a logarithm (or log) is defined as the power to which ten must be raised in order to equal the original number. Because most concentrations involve negative exponents (and most of us still find negative numbers difficult to think about), the pH function is defined as a negative logarithm. The pH, then, is the negative log of the hydronium or hydrogen ion concentration:

$$pH = -\log_{10}[H_3O^+] \text{ or } pH = -\log_{10}[H^+]$$

The pH function is more than a mathematical device. It is a very convenient way to keep track of the amount of hydronium ion in solution *without* using scientific notation. Also, because pH is based on a $\log_{10}$ function, we can easily compare pH values by factors of 10. A solution with a pH of 3.23 has 10 times more hydrogen ion per volume than a solution with a pH of 4.23.

When we have the pH, we sometimes must calculate the hydrogen ion concentration. This is done by "undoing" or performing the "inverse" of the log operation. The inverse operation of log is 10^x. Thus, for the function $pH = -\log_{10}[H_3O^+]$, the inverse function becomes

$$[H^+] = 10^{-pH}$$

We can also use a similar function on other variables. For example when applied to the concentration of hydroxide, we get $pOH = -\log_{10}[OH^-]$. When used with K_w, the value for the equilibrium constant for the ionization of water, we get $pK_w = -\log_{10}K_w$, and for the acid ionization constant K_a we obtain $pK_a = -\log_{10}K_a$.

In addition, we can derive an important relationship between the pH, pOH, and pK_w:

$$pOH = -\log[OH^-]$$

$$pK_w = -\log K_w = 13.996 \text{ (at } 25°C)$$

$$pK_w = -\log [OH^-][H^+]$$

$$= -\log [OH^-] - \log[H^+]$$

$$= pOH + pH$$

The pH scale allows us to quickly label a solution as **acidic** or **basic** relative to neutral water. For pure water at 25°C we can determine that the concentration of hydrogen ion is $1.0 \ 10^{-7}$ M. This gives a pH of 7.00. We say that any solution of water with a pH of 7.00 at 25°C is *neutral*—even if it is not pure. Water with more hydronium ion than neutral water, with $[H^+] > 1.0 \times 10^{-7}$, is said to be *acidic*. Acidic solutions at 25°C have a pH of less than 7. Water with less hydronium ion than neutral water, that is, with $[H^+] < 1.0 \times 10^{-7}$, is said to be **alkaline** or **basic.** Basic solutions at 25°C have a pH greater than 7.

If we have a solution in which a strong acid controls the amount of hydrogen ions, then it is relatively easy to calculate the value we expect for the pH. This will depend simply on the concentration of the acid, since every mole of the acid gives rise to one mole of hydrogen ions.

If we have a solution whose pH is controlled by a strong base, then the $[OH^-]$ concentration is set by the strong base. We can calculate an expected pH from

$$pK_w = pOH + pH$$

$$pH = pK_w - pOH$$

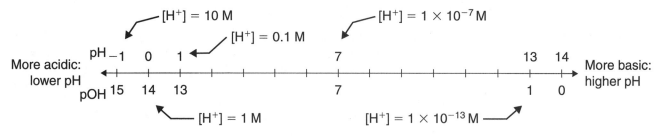

FIGURE E-1 THE pH SCALE WITH REPRESENTATIVE VALUES FOR [H⁺]

At 25°C the value of pK_w is 13.996. Therefore if a strong base is controlling the pH, then $pH = 13.996 - pOH$.

Some values for [H⁺] and pH are shown on the number line in Figure E-1. Notice something very important. Because of the nature of the pH function, the pH is *smallest* when [H⁺] is greatest. Thus, a solution with a smaller value for pH is more acidic than one with a larger value for pH.

pH of a Buffer

The equilibrium analysis of buffer concentrations can be done using the K_a expression given a known concentration of an acid and its conjugate base. If we are certain that neither A⁻ nor HA reacts very much (less than 3%, a good assumption if [A⁻] and [HA] are greater than 0.050 M and the pH is between 3 and 11), then we can use the Henderson–Hasselbalch equation to get the pH of a buffer:

$$pH = pK_a + \log([A^-]/[HA])$$

The Henderson–Hasselbalch equation is very helpful in calculating an expected pH for a buffer. If we know the pK_a of a weak acid and the ratio of the conjugate base and the conjugate acid in the buffer, then we can calculate the pH.

The equation also illustrates how buffers resist changes in pH. For example, if we have 0.100 L of a buffer that contains 0.025 M [HA] and 0.050 M [A⁻], then the ratio [A⁻]/[HA] is 2.0. If the pK_a of HA is 4.00, then the pH will be

$$pH = 4.00 + \log(2.0) = 4.00 + 0.30 = 4.30$$

If we add water to the buffer to increase the volume to 0.120 L, we lower the concentrations to 0.020 M [HA] and 0.040 M [A⁻], then we *still* have the same [A⁻]/[HA] ratio of 2.0. The pH should remain at 4.30.

Next, consider what happens when some strong acid is added to the buffer. In that case, some of the weak base is converted to the weak acid:

$$A^- + HCl \rightarrow Cl^- + HA$$

This means that the value for [A⁻] decreases while the value for [HA] increases. For example, let's say we have another 0.100 L of our buffer that contains 0.025 M [HA] and 0.050 M [A⁻]. If we add 0.020 L of 0.025 M HCl to the 0.100 L of a buffer,

1. We add $0.020 \text{ L} \times 0.025 \text{ M} = 5.0 \times 10^{-4}$ mol of HCl. We also increase the volume to 0.120 L (assuming volumes are additive).
2. The original buffer had $0.100 \text{ L} \times 0.050 \text{ M} = 5.0 \times 10^{-3}$ mol of A⁻. We will lose some of this in reaction with the HCl, so we are left with 4.5×10^{-3} mol of A⁻. The concentration of A⁻ will now be 4.5×10^{-3} mol/0.120 L = 0.038 M.

3. The original buffer had 0.100 L × 0.025 M = 2.5 × 10^{-3} mol of HA. We will increase the amount of HA through reaction with the HCl, so we wind up with 3.0 × 10^{-3} mol of HA. The concentration of HA will now be 3.0 × 10^{-3} mol/0.120 L = 0.025 M.

4. The ratio [A$^-$]/[HA] is now 0.038/0.021 = 1.5, so the pH is calculated to be

$$pH = 4.00 + \log (1.5) = 4.00 + 0.18 = 4.18$$

Despite the addition of 20. mL of strong acid, the pH changes very little. A similar control is exerted when strong base, for example NaOH, is added to a buffer.

Titration and the Analysis of Buffer Capacity

Buffers are important because they control pH, but they do have limits. At a certain point in the addition of acid or of base to the buffer solution, the contents of the buffer will be overwhelmed. The point where that occurs in this experiment can be used to determine the concentration of the components of the buffer.

When the addition of an acid or a base overwhelms the buffer, then the pH changes rapidly with the addition of a small amount of acidic or basic solution from the buret. This is known as the *end point* of the reaction (even though you should collect data beyond that point!). The end point should, if you were careful, be close to the *equivalence point* of the reaction. The equivalence point is the point when the amount of added acid is exactly equal to the amount of the base in the original solution, or vice versa. At the equivalence point, assuming that the acid and the base both donate or accept just one hydrogen ion, the following is true:

When adding acid to the buffer	When adding base to the buffer
moles base in buffer =	moles acid in buffer =
moles added acid	moles added base
$c_{base}V_{base} = c_{acid}V_{acid}$	$c_{acid}V_{acid} = c_{base}V_{base}$
$c_bV_b = c_aV_a$	$c_aV_a = c_bV_b$

These equations are very helpful, but they can be used in the wrong situation. In particular, they apply *only at the equivalence point* and *only for titrations where the stoichiometry of acid to base is 1:1.*

For both of these equations you know the volume (V, in L) and the concentration (c, in mol/L) of the added solution, and the volume of the buffer. It is then possible to solve for the concentration of the acid or the base in the buffer.

 Caution: Handling solutions of acids and bases requires special precautions. Though they do not stain clothing, a few drops of either can harm fabric, something that is only apparent after laundering. The solutions of NaOH that you are handling pose a particular risk to the eyes.

PROCEDURE

This experiment should be done by students individually. If needed, students may share burets, but each student must record a separate set of data for each part, to ensure familiarity with the pH meter and the burets.

Each student will record four tables of data in this laboratory:

1. Addition of strong base to deionized water.
2. Addition of strong acid to deionized water.

3. Addition of strong acid to a known concentration of a buffer solution.
4. Addition of strong base to a known concentration of a buffer solution.

It is essential that the notebook show both the values of any buret readings and the pH reading. All buret readings should be made to ±0.02 mL.

To ease the collection of data and later calculations, we suggest the layout of headings given in Table E-1 when working with deionized water. Don't forget to include a table entry for the solution *before* addition of the acid or the base.

TABLE E-1

Buret Reading	pH	Color	V (added)

When adding strong base or acid to a buffer solution, we suggest the headings in Table E-2. Again, don't forget to include a table entry for the solution *before* addition of the acid or the base. Also, record the initial concentration of the conjugate acid and the conjugate base, and calculate how many moles of each of them is present before the experiment begins.

TABLE E-2

INITIAL DATA:

V (buffer): _____   V (indicator): _____
(weak acid): _____   Moles of weak acid: _____
(weak base): _____   Moles of weak base: _____

Buret Reading	pH	Color

Part I: Preparation of Deionized and Degassed Water

In this experiment you will need a supply of pure water. Part of this has been done for you by a filtration system that removes any ions from solution. However, carbon dioxide from the air can dissolve in water to make carbonic acid, and this can compromise your results. Therefore you need to drive all the carbon dioxide from the water by boiling.

Obtain about 100 mL of deionized water in a clean beaker. Place it on a hot plate and bring the water to boiling; keep it boiling for 10 min. Remove the beaker and let it cool, covered by a watch glass to protect it from the room air.

Part II: Calibration of pH Meter

Obtain a pH meter from your instructor. Be certain to handle the pH probe carefully; its glass tip can be very fragile. Also obtain approximately 15 mL of a standardization solution in a clean, dry beaker. This solution is prepared to have a known pH value; you will adjust your pH meter to the value of the standard.

If necessary, plug in the meter and set it to measure pH. Turn on the meter and allow it to warm up for least 3 min. Gently rinse the tip of the probe with a stream of deionized water, then place the probe in the standardization solution. After allowing the probe and the solution to reach the same temperature, adjust the meter until it reads the correct value for the pH.

Part III: Addition of Strong Acid and Base to Deionized/Degassed Water

A pair of students may share two burets. One should clean and pre-rinse a buret with a solution of the standard strong acid. The other should do the same with standard strong base.

Use a clean, dry pipet to transfer 25.00 mL of your freshly degassed water into a clean, dry beaker equipped with a small magnetic stir bar. Add five small drops of bromocresol green indicator, then turn the stirrer on. Adjust so that the stirring is smooth with no spattering. Place the pH meter probe in the solution, avoiding any contact between the probe and the stirring bar. Record the pH and the color in your notebook. Use the format of Table E-1.

Add strong acid to the solution in small (0.25-mL) increments until the pH drops below 2.00. It is not necessary to add *exactly* that volume in each case, but you should aim for an increment around that value. In every case, record the actual buret reading, pH, and color of the solution at each point.

Next, prepare a new beaker with 25.00 mL of degassed water and five drops of bromocresol green. Carry out the experiment with the addition of 0.25-mL increments of the strong base. Continue until the pH rises above 11.00.

Part IV: Addition of Strong Acid and Base to Standard Buffer

In this part of the experiment, you will record the pH changes that occur when strong acid and strong base are added to a buffer solution containing sodium propionate and propionic acid.

Place 25.00 mL of the buffer and five drops of bromocresol green in a beaker, with a stirring bar. Begin stirring, then place the pH probe into the solution. Add the strong acid in approximately 0.50-mL increments. Again, it is not necessary to add *exactly* that volume in each case, but you should aim for an increment around that value. In any event, record the actual buret reading, pH, and color of the solution at each point. Use the format of Table E-2. Continue until the pH drops below 2.00.

Repeat this same procedure with a fresh sample of buffer and a clean beaker; use strong base this time and continue adding increments until the pH rises above 11.00.

REPORT

You have four parts to the procedure today. All sets of data should be graphed with volume of HCl or NaOH on the *x*-axis and the pH on the *y* axis.

Your interpretation of the results should be done separately for Parts III and IV of the lab.

Addition of the Acid and the Base to the Degassed Water

Before you prepare your report, you should finish your tables in the format of Table E-1 by adding an additional three columns (Table E-3). Remember that the number of moles added is equal to the concentration of the acid or the base (in moles per liter) times the volume (in liters) that is added. The concentration of added reagent is just the number of moles added divided by the total volume at that point.

TABLE E-3

ANALYSIS OF ADDITION OF STRONG ACID AND BASE TO DEGASSED WATER		
V (total)	Moles Added	Concentration of Added Reagent

Prepare a graph with both the experimental pH values and the calculated values at each value of V (added). All sets of data should be graphed with the volume of HCl or NaOH solution on the x-axis and the value of the pH on the y-axis

Addition of the Acid and the Base to the Buffer

Before you prepare your report, you should finish Table E-4. Remember that the number of moles added is equal to the concentration of the acid or the base (in moles per liter) times the volume (in liters) that is added. This added acid or base will react with the buffer. The number of moles of weak acid and of weak base change, and the ratio of these (last column) can be used to calculate a pH value using the Henderson–Hasselbalch equation.

TABLE E-4

ANALYSIS OF ADDITION OF STRONG ACID AND BASE TO BUFFER				
V (added)	Moles Added	Moles of Weak Acid (n_{acid})	Moles of weak Base (n_{base})	*Before equivalence:* $\frac{n_{base}}{n_{acid}}$ *After equivalence:* concentration of excess strong acid or base

For example, suppose a student starts with 25.00 mL of a solution that has 0.028 M formic acid (the conjugate acid) and 0.018 M sodium formate (the conjugate base). An additional 0.25 mL of indicator is added. At the start of the reaction,

"moles of weak acid" = 0.02500 L × 0.028 mol L^{-1} = 0.00070 mol
"moles of weak base" = 0.02500 L × 0.018 mol L^{-1} = 0.00045 mol

At a later point in the titration, a total of 5.45 mL of 0.0483 M sodium hydroxide is added. This means that

"moles added" is 0.00545 L × 0.0483 mol L^{-1} = 0.000263 mol of NaOH
"moles of weak acid" = 0.00070 − 0.000263 = 0.00044 mol of weak acid
"moles of weak base" = 0.00045 + 0.000263 = 0.00071 mol of weak base
"n_{base}/n_{acid}" = 0.00071/0.00044 = 1.6.

This is then used in the Henderson–Hasselbalch equation. With a pK_a for formic acid equal to 3.74, we get pH = 3.74 + log (1.6) = 3.94.

This method is used to calculate the pH values before the equivalence point.

Once the equivalence point is reached, the Henderson–Hasselbalch equation no longer applies, because either the weak acid or the weak base is used up. Then, there will be a certain amount of unreacted strong acid or strong base. You may continue to work with Table E-4, but now you can ignore the columns for n_{acid} and n_{base}. Instead, use the last column to calculate the moles of unreacted strong acid or base, then convert that to the calculated pH. This is the same as the calculation you did for the pH curve for Part III.

Prepare a graph with both the experimental pH values at each value of V (added) and the calculated values. Compare the results you obtained experimentally with the calculated values. Are there systematic differences between the two curves? How accurate do you expect a calculated pH to be, in general?

Determination of Total Acid and Base Strength

The addition of the strong acid and the strong base to the buffer initially creates a new buffer, with slightly different pH. Depending on the amount of the weak acid and the weak base present, this buffering capacity will eventually be overwhelmed at the equivalence point. This should be apparent from the experimental data curve, in the form of a much larger change in the pH. The volume of strong acid or strong base added to reach equivalence can be used to determine the strength of the buffer.

Present a calculation of the total number of moles of weak acid and weak base present, as determined from your titration curves. Is the buffer better at resisting changes when acids or bases are added? Why?

PREPARATION FOR NEXT WEEK

During the second experiment of this group you will use your knowledge of buffers to design a buffer. Before you leave this week, consult with your instructor to find out your assigned buffer concentration and pH. For next week you will each get an assignment as a member of a group; working together can help you see important similarities for your assignments.

Experiment 2
FOUNDATION LAB: DESIGN OF A BUFFER

Pre-Laboratory Assignment   **Due Before Lab Begins**

NAME: _____

Complete these exercises after reading the experiment but before coming to the laboratory to do it.

1. Use the Henderson–Hasselbalch equation to determine the ratio of acid to base in a formic acid–formate buffer with pH = 3.00.

2. If you need to prepare 250.0 mL of a buffer that has a total concentration of formic acid + formate of 0.030 M, then how many moles of each will you need to prepare the solution in question 1?

3. Discuss how to prepare the solution in question 2 starting from 0.100 M formic acid and 0.150 M NaOH.

4. Discuss how to prepare the solution in question 2 starting from 0.100 M formic acid and 0.150 M sodium formate.

5. Mixing acids and bases rapidly can be very dangerous. Before pouring any acid or base in a disposal unit, what can be done to neutralize the solution?

Experiment 2
FOUNDATION LAB: DESIGN OF A BUFFER

BACKGROUND

You now have experience in the action of buffers: they will resist the addition of acids and bases by shifting the relative amounts of the buffer components to compensate. Of course, buffers do not have infinite capacity to resist changes.

Here are some important elements of the buffer design:

1. The buffer must have the correct target pH.
2. It must be easy to prepare the buffer from convenient, pure components.
3. The buffer must have a reasonable capacity to resist the addition of strong acids and bases.
4. In some cases, the buffer must be prepared with the correct counterions.

In all buffers, three measurements are important. First is the pH of the solution. In the previous experiment, we noted the use of the Henderson–Hasselbalch equation to get the pH of a buffer:

$$pH = pK_a + \log ([A^-]/[HA])$$

The second number that is important is the *total* concentration of the weak acid and the weak base. This is used to indicate the concentration of the "buffer," although it is important to remember that we never have either component at this concentration.

$$[buffer] = [HA] + [A^-]$$

The third number, which is often not noted, is the ratio of the concentration of the weak acid to the concentration of the weak base. This ratio gives us some idea of the stability of the buffer. The most stable buffers have a ratio of 1:1. Generally, chemists like to have buffers with a ratio between 10:1 to 1:10, although, as was mentioned in the scenario, the blood buffer has a "precarious" 1:20 ratio of carbonic acid to hydrogen carbonate, yet it still keeps us alive.

Methods of Buffer Design and Preparation

Two methods of preparing buffers are described here. You will use both in this lab.

Method A. Prepare the buffer by adding some strong base to a weak acid. This converts some of the weak acid into its conjugate weak base, establishing the buffer. The pH is monitored to bring the value to the target; then water is added to make the volume the correct volume for the desired solution.

Example: Prepare 0.500 L of a pH = 4.00 solution with a total concentration of 0.00750 M using benzoic acid, $C_7H_8O_2$. Benzoic acid has a pK_a of 4.19. We will need 0.00375 mol of benzoic acid + benzoate in the buffer. Therefore, we take 0.00375 mol of benzoic acid (0.4580 g) and dissolve it in about 300 mL of deionized water. Place a clean pH meter probe, calibrated to a pH around 4, in the solution and add 2 M NaOH dropwise. Swirl gently to mix the solution well between drops, and continue until the pH is 4.00. Then, transfer the entire solution into a 500 mL volumetric flask and add water with mixing to bring the total volume to 500.00 mL.

For Method A to result in the desired buffer, it is essential we

- get enough moles of the benzoic acid to start, but we don't have to worry about the initial solution volume, as long as it is much less than the final volume.
- use a much more concentrated solution of sodium hydroxide, and we add it slowly so that we hit the target pH exactly (a calibrated pH meter is essential for this!).
- get all of the buffer components into the volumetric flask; it is important to rinse the beaker we used at the start with additional deionized water, then transfer it to the volumetric flask.

Method B. Mix appropriate amounts of a weak acid and a weak base, then dilute to the correct final volume.

Example: Prepare 0.250 L of a pH = 4.00 solution of formic acid with a total concentration of 0.0400 M. Formic acid has a pK_a of 3.74. Assume we have stock solutions with 0.100 M formic acid and 0.100 M sodium formate. Although it is conceptually simpler to say "mix the correct amount of the acid and the base," we have to do much more in the initial calculation.

1. *Determine the correct [HCOOH]/[HCOO⁻] ratio*. This requires us to rearrange the Henderson–Hasselbalch equation:

$$pH = pK_a + \log\left(\frac{[HCOO^-]}{[HCOOH]}\right)$$

$$pH - pK_a = +\log\left(\frac{[HCOO^-]}{[HCOOH]}\right)$$

$$\frac{[HCOO^-]}{[HCOOH]} = 10^{(pH-pK_a)} = 10^{(4.00-3.74)} = 10^{+0.26} = 1.82$$

This means that the concentration of formate should be 1.82 times that of the concentration of formic acid: [HCOO⁻] = 1.82 [HCOOH]

2. *Determine the actual concentrations of formate and formic acid needed to make the 0.0400 M buffer*. We know we must have the two concentrations added to the total buffer strength of 0.0400 M. Therefore,

$$[buffer] = [HCOOH] +]HCOO^-]$$

substituting and solving for [HCOOH]

$$0.0400 \text{ M} = [HCOOH] + 1.82[HCOOH]$$

$$0.0400 \text{ M} = 2.82[HCOOH]$$

$$0.0142 \text{ M} = [HCOOH]$$

Now that we know [HCOOH] we can solve for [HCOO⁻]

$$[HCOO^-] = 1.82[HCOOH] = 0.0258 \text{ M}$$

3. *Mix the correct mole and volume amounts of formic acid and a formate salt to give the desired concentration*. We have determined that if we made 1 L of solution we would need 0.0142 mol of formic acid and 0.0258 mol of formate. Since we are going to make 0.250 L of solution, we need to mix 0.00355 mol of formic acid and 0.00645 mol of formate. We will get these from the stock solutions of 0.100 M in each component:

Volume of formic acid: $0.00355 \text{ mol HCOOH} \times \dfrac{1 \text{ L}}{0.100 \text{ mol}} = 0.0355 \text{ L formic acid}$

Volume of formate acid: $0.00645 \text{ mol HCOO}^- \times \dfrac{1 \text{ L}}{0.100 \text{ mol}} = 0.0645 \text{ L formate}$

Thus, we mix 35.5 mL of formic acid solution and 64.5 mL of sodium formate solution in a 250.00-mL volumetric flask, then add water with mixing to give a total solution volume of 250.00 mL.

 Caution: Many different solutions will be in use today. If one disposes of an acid or a base in the wrong manner, then mixing of acids and bases may occur in a very short time period. This has the potential to release a large amount of heat. Therefore, label solutions carefully and dispose of them in the proper container. If necessary, neutralize both acids and bases with sodium hydrogen carbonate or a similar reagent before putting them into a common waste receptacle.

PROCEDURE

If possible, groups should be formed and assignments made during the *first* week of the experiment group. Students should then complete Part II of this procedure before coming to lab for the actual experiment.

Part I: Formation of Groups

This laboratory will be carried out by individuals making buffers in two different ways. Each student will then give his or her solution to another student, who will determine if the pH and the concentration is correct.

Part II: Designing and Preparing a Buffer

You will be assigned a buffering agent (ammonium–ammonia, lactic acid–lactate, acetic acid–acetate), a total buffer concentration, and a target pH. Your partners will have different buffers to examine, but there will be similarities. Work together to design your buffers, making note of when your calculations are the same and when they are different.

You should prepare 100.00 mL of buffer at the target pH in two ways:

1. Mix the weak acid with NaOH to make the correct final pH (method A). Dilute to the correct final volume.
2. Mix the weak acid and the weak base (method B). Dilute to the correct final volume.

Part III: Testing the Buffer

When you have prepared your buffer solutions, your partner will test them against the design. The test of each will have three parts:

1. Determine the pH of the buffer using a pH meter.
2. Titrate 25.00 mL of the buffer with NaOH, using a pH meter to follow the changes. Determine the number of moles of the weak acid in the buffer by determining the number of moles of NaOH required to overwhelm the buffer. Use the number of moles of weak acid to determine the concentration of the weak acid.
3. Titrate 25.00 mL of the buffer with HCl, using a pH meter to follow the changes. Determine the number of moles of the weak base in the buffer by determining the number of moles of HCl required to overwhelm the buffer. Use the number of moles of weak base to determine the concentration of the weak base.

Experiment 3
APPLICATION LAB: CARDIAC ARREST! THE RESTORATION OF BUFFERS

Pre-Laboratory Assignment   **Due Before Lab Begins**

NAME: _____

Complete these exercises after reading the experiment but before coming to the laboratory to do it.

1. Assume that the human blood buffer includes, at any one point, 0.00080 M carbonic acid and 0.0016 M hydrogen carbonate. Determine the number of moles of each component present in a person with 7.00 L of blood. What is the ratio of carbonic acid to hydrogen carbonate in this case?

2. The carbonic acid in the blood can be converted into carbon dioxide, which is a gas. Determine the *volume* of gaseous carbon dioxide that can be formed from the blood in question 1. Assume P = 1 atm and T = 315 K.

3. Metabolic acidosis results in the addition of excess acid to blood. How many moles of acid must be added to the blood in question 1 to bring the carbonic acid/hydrogen carbonate ratio to the hazardous level of 1:10?

4. Assume that the acid in question 3 is lactic acid formed by the body in the anaerobic respiration of glucose.

   a. What mass of glucose must react to give the acid in question 3?

   b. If the resting human body consumes the equivalent of 350 g of glucose per day, how long in minutes will it take for this amount of glucose to be converted?

5. Although hydrogen phosphate and dihydrogen phosphate are weak acids and bases, they still must be disposed of properly. What pH should a solution be before it is considered "neutral"? How can you tell?

Experiment 3
APPLICATION LAB: CARDIAC ARREST! THE RESTORATION OF BUFFERS

BACKGROUND

As we outlined at the beginning of this experiment group, the buffering of the blood is important in maintaining health. If the pH of the blood drops below 6.9 or rises above 7.8, then death is likely. But even within the "safe" range, the optimum pH for many bodily processes is quite narrow. Therefore, slight disturbances to the pH can significantly impair normal processes. And, of course, these may be the same processes that are used to recover after a critical incident such as cardiac arrest.

When the heart stops, two important problems arise that threaten the blood pH. First, metabolic acidosis sets in. Normal metabolism requires oxygen, which converts a molecule such as glucose all the way to carbon dioxide:

$$C_6H_{12}O_6 + 6\ O_2 \rightarrow 6\ CO_2 + 6\ H_2O$$

When the blood stops providing oxygen to the tissues, however, the tissues can only carry out metabolism partway, through the breakdown of glucose into lactic acid:

$$C_6H_{12}O_6 \rightarrow 2\ H_3C\!\!-\!\!CH(OH)COOH$$

The lactic acid generated in this way is the same lactic acid that can build up in the muscles during anaerobic exercise. This causes a familiar aching and fatigue in the muscles, which of course encourages us to rest. However, during cardiac arrest this reaction occurs in *all* the tissues of the body, and they are not equipped to handle a lactic acid buildup. So they dump it, right into the blood stream. The lactic acid then reacts with the hydrogen carbonate ion. This converts some of the blood's HCO_3^- to H_2CO_3 and alters the $H_2CO_3{:}HCO_3^-$ ratio.

$$C_3H_6O_3 + HCO_3^- \rightarrow H_2CO_3 + C_3H_5O_3^-$$

A second problem is that there is some carbon dioxide in the tissues that is formed by the last bits of available oxygen. This, too, is dumped into the blood, but it cannot be exhaled because the heart has stopped and blood no longer moves to the lungs. Even after the heart is restarted, there can be quite a delay before the excess CO_2 is removed.

Thus, it is a common protocol to quickly administer hydrogen carbonate to cardiac arrest patients. But emergency room personnel don't have the time to "titrate" the patient. They must administer a reasonable amount right away, then check to see if the treatment has worked.

The carbonic acid–hydrogen carbonate buffer is difficult to study in the general chemistry laboratory because it is too easy to lose carbon dioxide to the room atmosphere. Therefore, in this lab you will work with a buffer that behaves similarly to the carbonic acid buffer: the buffer formed between the acid dihydrogen phosphate, $H_2PO_4^-$ and the conjugate base hydrogen phosphate, HPO_4^{2-}.

There are three parts to this experiment:

1. As a group you will examine the capacity of normal solutions to resist changes in pH as HCl and NaOH are added. This will tell the total amount of weak base and weak acid in solution.

2. Your group will receive solutions with pH values that are too low for a healthy person. You will have to use a pH titration to determine how the different amounts

of hydrogen phosphate affect the pH. This will let you and your group develop a protocol to use on your patients.

3. Individually, you will receive a solution from a patient with acute acidosis following a cardiac arrest. You will have to determine the pH of the patient's blood, then calculate how much hydrogen phosphate to add to get a healthy pH. This has to be done all at once, just like in the emergency room. Your instructor will check the pH of your solution to see if your patient's blood has been restored to a healthy pH.

Caution: In this lab you are working near the normal pH of blood—which is close to pH = 7. Although this pH is considered a "neutral" solution in chemistry (at 25°C), your solutions may become much more acidic or basic. Use a small piece of pH paper to be sure that a solution is neutral before disposing of it.

PROCEDURE

Part I: Form Groups

The group will carry out the characterization of the behavior of the blood buffer as it is treated with strong and weak base. Each group will have four students, two to work on the buffer + NaOH and two to work on the buffer + NaOH and two to work on the buffer + HPO_4^{2-}.

The group should obtain about 500 mL of blood buffer. This will have a dangerously low pH—about 7.00. Use 25.00 mL of buffer in each titration.

Part II: Response of the "Blood" Buffer to Acid and Base

Each pair of students should carry out duplicate pH titrations. One pair should add a solution of NaOH and the other a solution of HPO_4^{2-}. Collect good data to answer the following questions:

1. How many moles of NaOH must be added to overwhelm the buffer?
2. What are the concentrations of $H_2PO_4^-$ and HPO_4^{2-} in the original solution?
3. For phosphate buffer, what is the ratio of $H_2PO_4^-$ to HPO_4^{2-} in the "healthy" pH range. (*Note:* because this buffer is not the same as the carbonic acid–hydrogen carbonate buffer, this number is not 1:20.)

Part III: The Treatment Protocol

Your group will be developing a protocol that allows you to answer the following critical care question:

A patient has a body mass between 75 and 225 lb. The patient's blood pH, due to the disruption of blood buffer, is now between 7.00 and 7.25. How many mL of bicarbonate must be added to bring the patient's blood pH to between 7.35 and 7.45?

We will not be able to use real patients or the volume of blood in real people, but we will simulate the problem with the following substitution:

A patient has a blood volume of between 75 and 225 mL. The patient's blood pH, due to the disruption of a $H_2PO_4^-$–HPO_4^{2-} buffer, is now between 7.00 and 7.25. How many mL of 0.050 M HPO_4^{2-} must be added to bring the patient's blood pH to between 7.35 and 7.45?

To answer these questions, your group will receive a sample with a pH that is too low for a healthy patient. By monitoring the effect of added hydrogen phosphate you will determine the amount of HPO_4^{2-} to add for your patient.

The question defines a range, just as a hospital team must handle a range of patients and a range of unhealthy pH values. The team in the hospital must adjust the amount of base to bring the pH to the correct value.

A good protocol should look like a grid: Put blood volumes across the top, and pH values down the side. Use increments of 25 mL in the volume and 0.05 in the pH units. In the boxes, enter the volume of hydrogen phosphate solution your team thinks it will need.

Part IV: Save Your Patient!

Go to your instructor and get an unknown slip. Your instructor will write down the time you got the slip. You have 30 min.

The slip will direct you to one of the patient solutions in the lab. Take the indicated volume from the stock bottle. Bring it to your bench, measure the pH, then add *all at once* the amount of HPO_4^{2-} you think you need. Mix well, and bring the solution to the instructor, who will measure the pH with a pH meter and let you know if you have succeeded.

EXPERIMENT GROUP F

MEASUREMENT OF CHEMICAL REACTION RATES: CLEAN UP WASTE WATER

Determining the contents of waste water is important to environmental safety.
Alan D. Carey/Photo Researchers.

PURPOSE The rate of a reaction and discovering inexpensive ways to increase that rate are important in a large number of chemical systems, from the simplest combustion process to the most complex biological transformation. Therefore, studying reaction rates, or reaction kinetics, is critical in all areas of chemistry and their application. This group of experiments will introduce you to the kinds of problems that chemical engineers face in their work in studying and using chemical reactions. There are three different methods used to follow reactions, but all involve the tactic of continuous monitoring.

In the first lab the method of spectrophotometry is used to characterize the formation of a colored product of a reaction. In the second lab you will monitor a reaction that consumes NaOH, a strong base. In the third lab you will monitor a reaction that generates oxygen gas.

SCHEDULE OF THE LABS

EXPERIMENT 1: Skill Building Lab: "Hydrolysis of an Ester"

1. Determine the reaction rate for the hydrolysis of *para*-nitrophenylacetate.
2. Examine the effect of added nitrogen-containing compounds on the rate.

EXPERIMENT 2: Foundation Lab: "Decomposition of t-Butyl Chloride in Base"

1. Develop a procedure for following the displacement of chloride by hydroxide.
2. Collect data on the concentration of hydroxide in the reaction.
3. Analyze reaction rates for different concentrations of 2-chloro-2-methylpropane (*t*-Butyl chloride) to derive the rate law for the reaction.

EXPERIMENT 3: Application Lab: "Decomposition of Hydrogen Peroxide"

1. Design a procedure for following the production of oxygen gas from decomposition of hydrogen peroxide.

2. Evaluate the rate and rate law of reaction with competing catalysts.

3. Analyze catalytic rates for engineering efficiency of the reaction.

SCENARIO Here is a dialogue that shows how an engineer, working with chemistry, can encounter the problem of reaction rates in industry.

"Oh, boy," your boss says, "the Feds are really starting to crack down on TBC (*tert*-butyl chloride) pollution. If we don't find a way to treat our exit streams, we're going to get fined to the annual tune of $20,000 per ton of TBC released."

This is a real problem, you realize, because the Acme Chemical company, and more specifically, the process stream you are in charge of, releases 3.5 tons of pollutant TBC per year (on a daily basis, that's 19 lb/day). It's diluted to very low levels before release, of course, well within current regulatory limits, but apparently the noose is being tightened.

"My raise, my promotion!" you think to yourself. "If I can't make Acme money with my process stream, I'm the one who's going to get in trouble."

Yet you begrudgingly admit that the present state of ecological awareness is not a bad thing for society. Current initiatives are in "green syntheses," which use environmentally benign substances in synthesis with reactions that generate little or no waste.

"Let's get down to specifics," your boss continues. "Remember the ten-gallon holding tank we're currently bypassing? Let's make it into a batch tank reactor. Shoot the TBC into that and react it with water, and we'll get *tert*-butyl alcohol. And you do know what we can do with the TBA, don't you?"

You reply immediately, "Isn't it a component of that gasoline additive MTBE? I think the stuff boosts engine combustion efficiency. If that's the case we can sell it to our sister company, the Acme Petroleum Company."

"You new college grads are great at seeing the big picture," your beaming boss admits. "Now I figure that if 90% of the TBC is converted into TBA, what's left over will be below the newest regulations after it's separated from TBA and diluted. For safety, we'll need to use water in the amount of one thousand times in excess of stoichiometry. Water's cheap, anyway, at $1.50 per thousand gallons. We can sell the TBA at $1400 a ton. It's going to cost us $100 an hour to run the doggoned reactor, though. All the power, safety, and personnel costs can eat us alive. That's why we bypassed the thing in the first place."

"Find out how many times a day we'll need to run the reactor, what the least amount of time (and cost) per run is, and the total annual net cost considering the sale of TBA. If it is possible to get the reaction to run any faster with base, let me know. That may allow us to take care of another problem we may have, with NaOH. Oh, by the way, I'll give you 10% of the difference between the cost of the fine and the cost of the clean-up as your next year's raise."

The chemical industry has provided many benefits to society, typically through the synthesis of new compounds with beneficial properties. But the "other side" of chemistry is the production of waste that must be discarded in an efficient and complete manner.

Oxidation reactions are an important class of chemical reactions. One of the best ways to carry out an oxidation reaction is with a chlorine derivative, such as chlorine bleach. However, this has the potential to produce a large number of waste products containing organochlorine groups. Organochlorine groups are implicated in a number of environmental problems, so green synthesis seeks to replace chlorine oxidants with nonchlorine oxidants.

One of the most popular alternative oxidants is hydrogen peroxide, H_2O_2. This produces water, not chlorine or chloride, as a byproduct. Hydrogen peroxide, though, still presents a waste disposal problem because it is a direct threat to many plants and animals. Therefore, though peroxide-based syntheses present less of a byproduct threat, engineers must still ensure that waste does not contain any unreacted hydrogen peroxide.

A final consideration for this scenario is the design of a system that will work on a waste effluent that will be discharged from a synthetic reaction. Such effluents are not collected: the unwanted materials must be removed as they flow out of a reaction system, but before they reenter the environment.

Chemical engineers are constantly facing challenges such as these. They must have a mastery of basic chemistry, and they must be able to design processes that are efficient and reliable. One component of the design is that the reactions must be continuous. You will get an experience with this in this group of experiments. While the reactions are done, in "batch mode" (that is, one at a time), you will get data about the progress of the reaction by continuously checking the change in one parameter of the system while not disturbing the reaction itself.

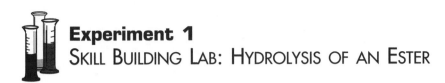

Experiment 1
SKILL BUILDING LAB: HYDROLYSIS OF AN ESTER

Pre-Laboratory Assignment   **Due Before Lab Begins**

NAME: _____

Complete these exercises after reading the experiment but before coming to the laboratory to do it.

1. The reaction you are studying generates acetic acid as one of the products. This could affect the pH of the solution. What precaution is taken against this?

2. Sketch graphs for the dependence of the reaction absorbance A with time for a reaction that is (a) zero order in PNA and (b) first order in PNA.

3. A buffer is prepared from 0.30 M phosphate stock solution, and 25.00 mL of this is diluted to a final volume of 100.00 mL. What is the concentration of the phosphate in this solution?

4. The buffer for this reaction contains dihydrogen phosphate and hydrogen phosphate. Dihydrogen phosphate has a K_a of 6.23×10^{-8}. At pH = 7.00, what is the ratio of [dihydrogen phosphate] to [hydrogen phosphate] in this solution?

5. What should be done to dispose of the reaction solutions in this experiment?

Experiment 1
SKILL BUILDING LAB: HYDROLYSIS OF AN ESTER

BACKGROUND

Hydrolysis Reactions

The reaction of a compound with water is very important in chemistry, especially in biological systems. In many cases, such a reaction results in a *splitting*, or *lysis* of the compound into two parts. Such "splitting-by-water" reactions are called *hydrolysis* reactions.*

Many different molecules contain a group of atoms that are called an ester, and these can be hydrolyzed by water to make a —COOH group (also called a carboxylic acid) and an HO— group (which, when the HO is attached to a C atom, are usually called alcohols).

$$\underset{R}{\overset{O}{\underset{}{C}}}\underset{O}{}R + H_2O \longrightarrow \underset{R}{\overset{O}{\underset{}{C}}}\underset{O}{}H + H\underset{O}{}R$$

This reaction is spontaneous for almost all esters. However, it can be very slow under typical conditions of temperature and pressure. It is much faster if there is a significant amount of a base in the solution, whether that is hydroxide ion or another base.

In this week's experiment, you will study the rate of this reaction using an ester that produces an alcohol, *para*-nitrophenol (PNP). The ester is called *para*-nitrophenylacetate, or PNA.

$$\text{[structure]} + H_2O \longrightarrow \text{[structure]} + \text{[structure]}$$

Under the conditions studied here, both the acetic acid and the *para*-nitrophenol lose hydrogen ions to make anions in steps that are fast relative to the hydrolysis of the ester itself. The PNP anion has a characteristic yellow color due to its absorption of light at 405 nm.

Caution: Neither *para*-nitrophenol nor *para*-nitrophenylacetate is listed as a carcinogen by the National Institutes of Health, but they have been found to cause chromosomal abnormalities. Therefore, they should be handled with extreme care. Proper gloves, careful washing, and disposal in designated waste containers are essential.

The phosphate buffers and catalyst solutions are potentially irritating. Thoroughly wash any skin areas that come into contact with these and clean up any spills immediately.

---

* There are other kinds of hydrolysis reactions, including the reaction of acids and bases with water.

Procedure

Part I: Formation of Groups

You will work in groups of two, with each pair of students studying one phosphate concentration and one catalyst. Later on, the group results will be pooled to get data about two different phosphate concentrations and several different catalysts.

Each group will have several different solutions in use at any one time. Carefully label all beakers and flasks.

Because of the dramatic effects of catalysis, be certain to wash all glassware carefully before using. Pipets should be used for one solution at a time; do not use the same pipet for different solutions without careful cleaning and prerinsing.

Part II: Preparation of Phosphate Buffer Solution (Group Work)

Each pair of students will examine the rate of the reaction at a different concentration of a phosphate buffer. The buffer contains both hydrogen phosphate and dihydrogen phosphate ions, with the sum of their concentrations equal to the amount of phosphate in the buffer. Thus, when we say that a solution is 0.10 M phosphate buffer we do not mean that the concentration of PO_4^{3-} is 0.10 M. We mean that the concentration of all phosphate-related substances and ions (phosphoric acid, dihydrogen phosphate, hydrogen phosphate, and phosphate ion itself) add to 0.10 M. The actual concentration of the components depends on the pH of the solution.

All experiments by a pair of students will be done with the same buffer, so all solutions must be prepared from the same buffer solution.

Obtain 25.00 mL of phosphate buffer (either 0.20 M or 0.40 M) in a 50- to 100-mL beaker. This will have a pH of about 6.50, and you have to adjust the pH to 7.00. Place a *calibrated* pH probe in the solution, then add 2 M KOH solution *dropwise* to bring the pH to 7.00. When you are done, transfer all of your solution to a clean 100.00-mL volumetric flask that has been rinsed twice with deionized water. Rinse the beaker with a few milliliters of deionized water, and add the rinsing to the volumetric flask. Finally, add deionized water to bring the total solution volume to 100.00 mL.

Part III: Determination of the Rate of Hydrolysis in Buffer (Group Work)

In this experiment you will measure the appearance of *para*-nitrophenol (PNP) in a solution. Standardize the spectrophotometer with a blank of the phosphate buffer only using the reference wavelength of 405 nm. You will then study the rate of hydrolysis in the buffer itself. This slow reaction will show the importance of the catalyst.

Obtain about 20 mL of PNA solution in a clean, dry beaker. This solution is stable as prepared, but it begins to react when it is put in a solution above pH = 6.5. Therefore, do not add PNA to any solution until you are ready to begin the measurement of the appearance of PNP.

You start the reaction by mixing a certain amount of buffer with a fixed amount of the PNA solution. The exact values will depend on the volume of the spectrophotometric cell you are using. If you are using cuvettes, then the total volume is about 2.80 mL. Use a Mohr pipet to add 2.30 mL of the buffer to a cuvette. When you are ready to begin the experiment, add 0.50 mL of the PNA.

To mix the solutions and start the reaction, cover the cuvette securely and invert it to mix the solutions well, or use a Pasteur pipet to quickly withdraw and reinject the solution in the cuvette several times.

Try to take the first absorbance measurement at 30 s after mixing, and then take additional measurements every 5 min for 30 min.

While this reaction is proceeding, one of the students in the pair can prepare the two solutions of the catalyst candidate needed for Part IV of the experiment.

Part IV: Determination of the Rate of Hydrolysis with a Catalyst

Your instructor will assign a catalyst to each group. Obtain 10 mL of the catalyst solution in a clean, dry beaker. Note the concentration of the catalyst.

Prepare two solutions of the catalyst in the buffer. For the first, transfer 2.50 mL of the catalyst and 2.50 mL of deionized water into a 25.00-mL volumetric flask, then dilute to the mark with your phosphate buffer. For the second, use 5.00 mL of the catalyst in a second volumetric flask, then dilute again with your buffer.

Prepare hydrolysis solutions as before: Use the *same* amount of buffer + catalyst and PNA solution as in Part III.

If a pair of students plans well, then all three samples—the phosphate buffer only and the first and second catalyst solution—can be studied at the same time by careful switching of the solutions in the spectrophotometer. All solutions should be measured every 5 min for 30 minutes.

Before discarding the solutions containing catalyst, it is necessary to measure A when all the PNA has reacted. To do this, add one drop of concentrated catalyst from the dropper bottle that will be in the lab. Mix the cuvette well, wait five minutes, and record A_{infinity}.

REPORT

Your data need to be analyzed for the increase in the concentration of a product PNP with time. However, you will not be able to determine the absolute concentration except by a multistep calculation. Instead, you can determine the rate of the reaction by looking at the absorbance values only.

If we have a reaction in which only one species, a product, is absorbing light, then we can follow the increase in the absorbance to determine the reaction rate. This assumes that the species obeys Beer's Law under the conditions of the experiment. In that case $A = \varepsilon cl$, where ε is a coefficient for the amount of light absorbed, l is the length of the cell, and c is the concentration of the substance. In this experiment, we do not determine c itself. Instead, we follow the relative amount of the product by comparing A at different times.

To get the reaction rate, we analyze the reaction using the first-order kinetics in the PNA. This means the rate = $k[\text{PNA}]$.

Using the appropriate integrated rate law, we get a dependence of the concentration of PNA on the time: $[\text{PNA}] = [\text{PNA}]_0 e^{-kt}$. At time $t = 0$, $[\text{PNA}] = [\text{PNA}]_0$ and at time $t = \infty$, $[\text{PNA}] = 0$.

In this reaction, PNP is directly produced from PNA in a 1:1 ratio. Therefore, the final concentration of PNP is equal to the initial concentration of PNA. At any time t, between the start and the finish,

$$[\text{PNP}] = [\text{PNA}]_0 - [\text{PNA}]$$

$$= [\text{PNA}]_0 - [\text{PNA}]_0 e^{-kt}$$

$$= [\text{PNA}]_0 \left(1 - e^{-kt}\right)$$

$$= [\text{PNP}]_{\text{infinity}} \left(1 - e^{-kt}\right)$$

Assuming that Beer's Law is obeyed, $[\text{PNP}] = A/\varepsilon l$. Therefore,

$$A_t/\varepsilon l = A_{\text{infinity}}/\varepsilon l(1 - e^{-kt})$$

$$A_t = A_{\text{infinity}}(1 - e^{-kt})$$

$$A_t/A_{\text{infinity}} = (1 - e^{-kt})$$

$$(A_t/A_{\text{infinity}}) - 1 = -e^{-kt}$$

$$1 - (A_t/A_{\text{infinity}}) = e^{-kt}$$

$$\ln[1 - (A_t/A_{\text{infinity}})] = -kt$$

If we plot $\ln[1 - (A_t/A_{\text{infinity}})]$ vs. t we will get a line with a slope equal to $-k$.

Using your data, prepare a table that has time for the first column, A in the second column, and $\ln[1 - (A_t/A_{\text{infinity}})]$ in the third column. Use these values to prepare graphs of A_t vs. t and $\ln(1 - A_t/A_{\text{infinity}})$ vs. t. Determine the slope m for this line and calculate k. To assess the effect of the catalyst, compare the graphs for different catalyst concentrations. The *relative* slopes of these graphs will provide the order of the reaction in the catalyst.

Experiment 2
FOUNDATION LAB: DECOMPOSITION OF *t*-BUTYL-CHLORIDE IN BASE

Pre-Laboratory Assignment   **Due Before Lab Begins**

NAME: _____

Complete these exercises after reading the experiment but before coming to the laboratory to do it.

1. Write the equations for the reaction of *t*-butyl chloride (2-chloro-2-methylpropane) and water to produce HCl and with hydroxide to give water and chloride.

2. a. If a solution has a pH of 10.25, what is the pOH and what is concentration of hydroxide ion?

   b. If this solution has a volume of 50.00 mL, how many moles of hydroxide ion are present?

   c. If 1.9×10^{-6} mol of NaOH are consumed in this solution, what will the hydroxide ion concentration, the pOH, and the pH be?

3. You observe a pH change of 1.0 unit. Can you determine the corresponding change in hydroxide ion concentration? If not, what additional information do you need?

4. If you conduct a reaction with 0.0850 g of *tert*-butyl chloride in 100.0 mL of solution, how many moles of *tert*-butyl chloride will decompose? How many moles of NaOH will be used? If you collect the data until at least 25% of the reaction has proceeded, what will be the final pH of the solution, if you begin with pH = 11.00?

5. Identify the flammable solutions and products in this procedure. What precautions must you take before handling any of these?

Experiment 2
FOUNDATION LAB: DECOMPOSITION OF *t*-BUTYL CHLORIDE IN BASE

BACKGROUND

In this experiment you will learn how to develop a kinetic method using the decomposition of an organochlorine substance, 2-chloro-2-methylpropane, also known as *tert*-butyl chloride.* Its formula is C_4H_9Cl.

The decomposition of C_4H_9Cl with water gives HCl and $C_4H_{10}O$, called 2-methylpropan-2-ol, or *tert*-butyl alcohol. *Tert*-butyl alcohol is a component of a new class of gasoline additives, the *tert*-butyl ethers, used to boost the combustion efficiency of gasoline.

The decomposition reaction of *tert*-butyl-chloride in water is

$$H_2O + C_4H_9Cl \rightarrow HCl + C_4H_{10}O$$

The easiest way to monitor the HCl produced is by its reaction with NaOH, a strong base:

$$HCl + NaOH \rightarrow H_2O + NaCl$$

The overall reaction in base is therefore

$$C_4H_9Cl + NaOH \rightarrow C_4H_{10}O + NaCl$$

This reaction can be monitored with a pH meter. The pH meter reports how much hydronium ion is in solution, because $pH = -\log_{10}[H_3O^+]$. Since we are measuring the disappearance of hydroxide, it is easier to convert pH to pOH, and then to get the hydroxide ion concentration.

$$pOH = 14.00 - pH \text{ (at 25°C)} \qquad [OH^-] = 10^{-pOH} = 10^{pH-14.00}$$

 Caution: The solutions of 2-chloro-2-methyl-propane used in this experiment are in iso-propyl alcohol. This combination is very flammable; hence, no open flames or other ignition sources should be allowed in the lab during this experiment.

The sodium hydroxide solutions you will use are caustic. Proper eye protection is very important.

PROCEDURE

Part I: Formation of Groups

This experiment should be done by pairs of students. They should each prepare a notebook for the kinetic data that will be produced in this experiment.

Part II: Monitoring pH in a Continuous System

It has been suggested that this reaction can be monitored by consumption of NaOH. How easy is it to do this accurately? To find this out, carry out a test in which 0.0050

---

* The mode of attachment of the chlorine to the carbon in *tert*-butyl chloride is very different from the structure found in toxic organochlorine compounds. This compound has no identified toxicity.

M HCl is added from a buret into a beaker of hydroxide in water. This must be done carefully, because simply adding the HCl to the top of the NaOH solution may not mix the solutions.

Place the pH meter in a beaker with 50. mL of NaOH with a pH between 11.20 and 11.50. Add a drop of phenolphthalein indicator. Monitor the pH change as 25 mL of 0.0050 M HCl solution is added to the NaOH. Do this without stirring or swirling the mixture. Then do it with stirring with a magnetic stir bar. As part of your report, you will have to justify your experimental setup by referring to what you observe here. Collect data here to show that you can reliably measure pH as a function of time. Note the pH when the color of the phenolphthalein changes.

Part III: Measurement of Decomposition of *tert*-butyl Chloride

Obtain a vial of 2.0% *tert*-butyl chloride in isopropyl alcohol from the supply in the fume hood. Be very careful not to let this contact any water. Make sure the beaker or flask you use to transport the solution is clean and dry, and do the same for any pipets. This flammable mixture can be safely handled in the lab if no flames or other ignition sources are present. Practice pipetting small (2.00-mL) amounts of this mixture using a *dry* 2- or 5-mL Mohr pipet.

> **NOTE:** It is very easy to contaminate the *tert*-butyl chloride with water. Take a portion of the stock solution in a well-dried beaker or small Erlenmeyer flask for your group to use.

You want to use enough NaOH so that change in the color of the indicator occurs at least 2 min after the *tert*-butyl chloride solution is added. To determine how much NaOH solution to use, add one or two drops of phenolphthalein to beakers containing 25, 35, 40, and 45 mL of the NaOH solution with lower pH. Then, add 2.00 mL of the *tert*-butyl chloride solution to each beaker, swirl for 5 s. Note the time it takes for the phenolphthalein to change color. You should do your kinetic runs on the volume of NaOH that gives the color change in a range of 2 to 5 min.

Transfer the relevant amount of NaOH into a beaker. Add a drop of phenolphthalein and the pH-meter probe. Add the 2.00 mL of the *tert*-butyl chloride solution, swirl for 5 s to mix, and begin collecting data. Collect data until the pH has reached a level at least 2 pH units below the initial value. This may take several minutes.

Part IV: Monitoring at a Different pH: Determination of the Reaction Order in Hydroxide

Hydroxide is a possible reactant in this system. Although determining the reaction order can be done by analyzing the concentration of hydroxide from Part III, it is easier to look at the rate of the reaction at a different initial concentration of hydroxide. Using the higher-pH NaOH solution in the lab, carry out a procedure to check if the reaction rate depends on hydroxide.

CALCULATIONS

Kinetics experiments are calculation intensive because many measurements must be processed. You should present a data table that contains entries for time, pH, $[H_3O^+]$, $[OH^-]$, and $[C_4H_9Cl]$. Note that the change in the concentration of the hydroxide is equal to the change in the concentration of the C_4H_9Cl. Also evaluate $\ln [C_4H_9Cl]$ and $1/[C_4H_9Cl]$. Plot all three functions of $[C_4H_9Cl]$ to determine the order of the reaction and the rate constant k.

You can download a spreadsheet that will assist you in your calculations. It can be found under the "Decomposition Kinetics" listing at the WWW address available from your instructor.

You will need either Microsoft Excel™ or Corel Quattro Pro for Windows™ to use these files.

Results and Conclusions

Your results should present an analysis of the characteristic plots of the kinetic runs.

For your conclusions, you should report the method you developed. How does it work? Did it work well with your solutions? Finally, what can you tell about how your method overcomes any possible problems in this system?

QUESTIONS TO ANSWER IN YOUR LAB REPORT

1. For what range of concentrations is your kinetic method best?
2. Imagine that you are designing a system that will accomplish the conversion of waste *tert*-butyl chloride into the more benign *tert*-butyl alcohol. Determine the time it will take for a solution with 0.0100 M *tert*-butyl chloride to react to give a solution with 1.0×10^{-4} M *tert*-butyl chloride.
3. If you must guarantee that the *tert*-butyl chloride waste from a system is destroyed in a certain amount of time, will the initial concentration of the *tert*-butyl chloride matter? Why or why not?

Scenario Questions to Answer in Your Lab Report

1. Calculate the yearly sales of TBA, assuming a 90% conversion of the number of *moles* of TBC to TBA. The price of TBA is $1400/ton.
2. Calculate the amount of water needed for the industrial reactor system, if the number of *moles* of water is to be 1000 times the number of moles of TBC. Calculate the yearly cost of the water at the price of $1.50 per 1000 gal.
3. Convert the number of moles of TBC and water reactants into volumes. The density of TBC is 0.9 g per mL. Calculate how many 100-gal reactors per day need to be run (at a 10:1 molar ratio of H_2O to TBC) in order to process the TBC.

Experiment 3
APPLICATION LAB: DECOMPOSITION OF HYDROGEN PEROXIDE

Pre-Laboratory Assignment   **Due Before Lab Begins**

NAME: _____

Complete these exercises after reading the experiment but before coming to the laboratory to do it.

1. A solution of hydrogen peroxide is usually sold as a 3% solution by weight. How many moles and grams of hydrogen peroxide are in 25.0 mL of such a solution if the density of the solution is 1.03 g mL^{-1}? .

2. What volume of oxygen gas is produced by the complete decomposition of the hydrogen peroxide in question 1? What assumptions must you make about the composition of air to carry out this calculation? Assume $P = 1$ atm and $T = 313$ K.

3. You must build your own apparatus to generate and collect O_2 gas from the decomposition of a hydrogen peroxide solution and then measure the volume of the O_2 gas as it is formed. Using common lab glassware and rubber tubing, sketch a device that will accomplish this.

4. What safety concern must be addressed in the design of the apparatus in this experiment?

Experiment 3
APPLICATION LAB: DECOMPOSITION OF HYDROGEN PEROXIDE

BACKGROUND

"Hey, here's another one for you," your boss says. "We've got the same hydrogen peroxide disposal problem as everybody else in this business, but our arch-rival, Megachemco, has just come out with a great new catalyst for catalytic peroxide decomposition. Speeds up the rate a million times over the uncatalyzed rate. Their reactor costs are now smaller than my daddy's income tax payments."

You recall that peroxide has become a popular alternative for toxic organochlorine compounds; these are used for the very lucrative oxidation reaction business. Acme made this switch itself a few years back. "More green synthesis," you note to yourself. In this case, the new, greener process apparently needs some major improvement.

You boss continues, "Since you did so well on the TBC project, you're my go-to guru for kinetics. If we don't come up with our own catalytic process, our business will tank real soon."

"Well all right," you respond, "what do we know about their catalyst?"

"Their patent claims that the stuff can decompose 95% of a 0.01 M solution of peroxide in a hundred-gallon reactor in twelve minutes. The catalyst has a molar mass around 850 g mol^{-1} and is only present is dilute amounts, 0.0001 M, so it must be very active. Our marketing people have placed calls to Megachemco; they tell us that the catalyst costs a fortune at $5,000 per kilogram."

"Then I've got an idea," you immediately brainstorm. "We don't necessarily need a catalyst that's more active than Mega's. If we can find one that's, say, half as active but costs 100 times less, we can add a lot more catalyst and get higher activity for the same amount of money."

"Hot dog," shouts your boss, pounding his fist on the desk, "I knew you'd think of something. You know, we have lots of potassium iodide ($100 per kilogram) and manganese oxide ($50 per kilogram) stockpiled around here. Why don't we try those first? Find out the particulars: for the same price per batch, how long will it take these other catalysts to decompose 95% of hydrogen peroxide under Mega's conditions? By what percent can we cut down their processing time?"

Now you are ready to design a kinetic analysis for the removal of hydrogen peroxide from a waste stream. The reaction you will employ is

$$2\ H_2O_2 \rightarrow 2\ H_2O + O_2$$

An efficient analytical method is to decompose the peroxide and measure the volume of gas produced.

Caution: During this lab you will study the decomposition of hydrogen peroxide solutions, which produces oxygen gas. Be careful never to allow this reaction to occur in a system that does not have a way to vent the gas. Otherwise, high pressures may result.

Solid manganese dioxide, though not listed as a toxin, is a fine powder that should be handled carefully to avoid breathing its dust.

PROCEDURE

In the report for this experiment you will be asked to calculate how to clean up waste water containing hydrogen peroxide. You will have to decompose all of the hydrogen peroxide in 10 min, using a method you devise. It will not, however, be enough for you to dump a large amount of a catalyst into the solution. You will have to justify your method on the basis of some efficiency measures using the data you collect today.

Part I: Formation of Groups

Each group of three will study two different catalysts. One is a solid, manganese(IV) oxide. The other is a soluble substance, potassium iodide. Think about how your calculations will require gas laws. What other measurements should you make?

Part II: Design Apparatus and Collect Data for H_2O_2 Decomposition

Work together to design a method for the monitoring of gas production from a solution, starting from the design in your prelabs. In doing this, you will have to guarantee that the gas is continuously transferred to the monitoring system: This may require you to carefully stir the mixture. Magnetic stirrers will *not* be available for this lab. That would unrealistically assume that a waste stream can be rapidly stirred.

Your monitoring system should take advantage of a device called a *eudiometer*, which is a long tube with volumetric markings. By filling it with water and then inverting it within a beaker, you will have a column of water that the evolved gas can displace. You can measure the volume with time continuously.

When you have a monitoring system designed, draw a sketch of the system in your notebook, then show it to your instructor. He or she may suggest improvements or point out flaws. Then, collect data for the decomposition of 50.00-mL portions of your peroxide.

You will need to carry out at least eight runs. They should be duplicates of both catalysts at two different "loadings" or concentrations of the catalysts.

REPORT

You will need to carry out calculations that describe the dependence of the reaction on $[H_2O_2]$. Your basic data will be a volume of gas in the eudiometer. Convert this into moles of oxygen produced by using the ideal gas law. Be careful to correct the pressure of the gas for the vapor pressure of water and for the difference in the height of the water columns inside and outside of the eudiometer. Then, study the rate law by graphing $[H_2O_2]$, $\ln[H_2O_2]$, and $1/[H_2O_2]$ vs. time. Finally, compare the rates at different catalyst loadings to get the dependence on the catalyst.

Perform the analyses to determine the rate expression for H_2O_2 and for each catalyst. For your analysis, use the simple expression:

$$\text{rate} = k''[W][H_2O_2]^x$$

with the goal of finding x and k'' for each catalyst.

QUESTIONS TO INCLUDE IN YOUR LAB REPORT

1. Report on the utility of the method you developed. How reproducible were the results? Did you notice lag times? How would particle breakup affect your measurements?
2. Once the rate law is determined, calculate the time required for 95% conversion of hydrogen peroxide under the conditions given in the scenario and using the

same dollar amount of catalyst for the [W] term. Compute the percent change in reactor time for each case.

3. Of course, you could add a huge amount of the catalyst, but cost is a factor in engineering processes. In your lab write-up, calculate the cost of the catalyst that you have used for this portion of the experiment.

4. Now answer the company's question: "What amount of catalyst, at what cost, is needed to decompose 95% of the H_2O_2 in a solution of 0.010 M H_2O_2 in a 100-gallon reactor within 12 minutes?"

5. "Now just wait a minute," you start to think, "In homogeneous catalysis the catalyst is intimately mixed with the reactants. If the catalyst is inexpensive, it can be left in with the products as an impurity. If it's expensive and needs to be recovered, an additional separation step must be added to the process." Along these lines, what advantages do solid catalysts present over homogeneous catalysts? How will this effect the economics of a process?

EXPERIMENT GROUP G

ELECTROCHEMICAL GLUCOSE MONITORING: CONSTRUCT AND USE YOUR OWN ELECTRODE

Many persons can carry out very good chemical measurements with simple devices.
David K. Crow/PhotoEdit.

PURPOSE   This is a three-week set of experiments on the theory and techniques used by chemical professionals in the field of electrochemistry. This experiment group examines two components of the electrochemical measurement: (1) the instrument used to make the appropriate measurements (generally a voltmeter) and (2) the electrodes that serve as the connection between the chemical system and the voltmeter. You will construct your own electrode and use it to determine the concentration of an unknown solution by two different methods: by measuring the cell potential difference directly and by monitoring the cell potential in a potentiometric titration. Finally, you will examine an electrolysis cell and compare its measurements to measurements made with an over-the-counter glucose monitor.

SCHEDULE OF THE LABS

EXPERIMENT 1: Skill Building Lab: " Electrochemical Measurement"

1. Construct a galvanic cell (group work).
2. Measure cell potential differences, $\Delta \mathscr{E}_{cell}$, for various galvanic cells (group work).
3. Determine $[Cu^{2+}]$ in a solution of unknown concentration using the Nernst equation (individual work).
4. Construct a working combination electrode (group work). Use the electrode to
   a. Measure $\Delta \mathscr{E}_{cell}$ for standard solutions of Ag^+ needed to prepare a calibration curve of $\Delta \mathscr{E}_{cell}$ versus log $[Ag^+]$ (group work).
   b. Determine the concentration of Ag^+ in an unknown solution by comparison with the calibration curve (individual work).

EXPERIMENT 2: Foundation Lab: "Potentiometric Titration"

1. Use the self-constructed electrode from Experiment 1 in a potentiometric titration of a chloride solution (known concentration) with a standard silver nitrate solution (group work).

2. Use the self-constructed electrode from Experiment 1 in a potentiometric titration of a chloride solution (unknown concentration) with a standard silver nitrate solution (individual work).

EXPERIMENT 3: Application Lab: "The Glucose Biosensor"

1. Examine the effect of the following factors on current in an electrolysis cell (individual work).
   a. electrode distance (individual work).
   b. electrode position (individual and team work).
   c. concentration changes (individual work).

2. Examine measurements made with a glucose test kit by
   a. Preparing the monitor (individual work).
   b. Preparing standard solutions (individual work).
   c. Preparing a calibration curve and unknown sample measurement (individual work).

SCENARIO A child has just been admitted to the emergency room at your hospital. Her symptoms include confusion, nausea, clammy skin, glazed eyes, and general lethargy. These symptoms, together with a family history of diabetes, suggest the patient is suffering from a glucose imbalance. Along with other initial testing, the doctor orders a blood glucose test for this patient.

The average person has an amount of blood glucose that falls within a particular range called the "normal" range. If the glucose level is too high (a condition called *hyper*glycemia), severe tissue complications may result. These patients are given a course of treatment to *reduce* their blood glucose levels to within the normal range. On the other hand, when the blood glucose level is too low (*hypo*glycemia), a person may lose consciousness or even lapse into a coma. These patients must be given a course of treatment tailored to *increase* their blood glucose levels. Either condition, if left untreated, can be fatal. It is important, therefore, that the correct blood glucose level be determined so that the patient may receive the proper treatment.

Glucose testing kits, which are commercially available at all major drug stores, allow individuals with abnormal blood sugar levels to monitor the increase and decrease of their blood glucose levels throughout the day. This "on the spot" testing helps them to maintain proper levels of blood glucose by giving them the information they need in a timely fashion. With this information, they can act to counteract any glucose imbalance as it occurs. This ability to direct their own health program allows them to live a normal life, usually within the bounds of a special diet.

Many over-the-counter glucose testing kits rely on electrochemical principles for their operation. A small current is produced when a droplet of blood is placed on the test strip. The blood dissolves some dry chemicals on the test strip and completes the circuit between two miniature electrodes. Each new set of test strips is calibrated with standard glucose solutions, and the current produced during the self-testing is "read" as glucose concentrations. As background for the glucose testing, this module introduces basic electrochemistry concepts and techniques.

Experiment 1
SKILL BUILDING LAB: ELECTROCHEMICAL MEASUREMENT

Pre-Laboratory Assignment   **Due Before Lab Begins**

NAME:_____

Complete these exercises after reading the experiment but before coming to the laboratory to do it.

1. An electrochemical cell consists of the half-cells, $Cu^{2+}|Cu$ and $Ag^+|Ag$.

   a. Write half-reactions for the reduction of Cu^{2+} to Cu and Ag^+ to Ag. Include the standard reduction potentials, $\mathscr{E}°$.

   b. Write the half-reactions that occur at the anode and at the cathode. Label these.

   c. Write the balanced equation for this reaction.

   d. How many electrons are transferred during this reaction?

   e. Write the reaction quotient, Q, for this reaction.

2. Sketch and label an electrochemical cell for $Cu^{2+}|Cu$ and $Ag^+|Ag$. Show the direction of flow of electrons in the external circuit for the spontaneous reaction.

3. Use the Nernst equation to calculate $[Ag^+]$ when $[Cu^{2+}] = 0.012$ M and $\Delta\mathscr{E}_{cell} = 0.410$ V.

4. Suppose that you have five solutions containing the metal ions of Fe, Cu, Ag, Pb, and Zn. List all possible metal ion pairs.

5. Evaluate $RT/\mathcal{F}$ when $R = 8.315$ J K^{-1} mol^{-1}, $T = 298.15$ K, and Faraday's constant, $\mathcal{F} = 96,485$ C mol^{-1}.

6. What is the purpose of a calibration curve?

7. You have a stock solution labeled 0.200 M silver nitrate. What volume of stock solution is needed to make up the following dilutions?

   a. 15.0 mL of 0.075 M silver nitrate

   b. 15.0 mL of 0.038 M silver nitrate

   c. 15.0 mL of 0.094 M silver nitrate

8. List two safety precautions that are important to follow when using electricity in an experiment.

Experiment 1
SKILL BUILDING LAB: ELECTROCHEMICAL MEASUREMENT

BACKGROUND

Electrochemistry uses an electrical signal such as voltage or current to measure chemical amounts (moles) or concentrations (moles per liter). One type of electrochemical measurement uses a spontaneous oxidation–reduction reaction (redox) to generate an electrical signal that is measured by a voltmeter and recorded in volts. This is the basis of a galvanic cell. The second type of electrochemical measurement requires an external source of electricity to produce a redox reaction. This is an electrolysis cell. The reaction that occurs in an electrolysis cell is not a spontaneous reaction.

A *spontaneous* reaction is one that occurs without continuous outside help. As with other thermodynamic processes, if a reaction is favored in the forward direction, then it is not favored in the reverse direction. Thermodynamically, spontaneous reactions are favored; nonspontaneous reactions are not favored. Spontaneity reflects the energy of the system. For electrochemical reactions, we can associate the change in the energy of the system with the driving force of the flow of electrons.

When properly configured, electrons in a redox reaction will flow through an external wire, and we can use a device to measure their electrical potential, or voltage. A spontaneous redox reaction will have a positive cell potential difference, $\Delta\mathscr{E} > 0$. A negative cell potential difference, $\Delta\mathscr{E} < 0$, indicates a nonspontaneous reaction.

Oxidation-Reduction Reactions

A redox reaction is the simultaneous occurrence of two components or half-reactions: one component is the oxidation process and the other is the reduction process. Oxidation occurs when a chemical species loses or gives up electrons to another chemical species. Reduction occurs when a chemical species receives or gains electrons. The oxidation process provides the electrons necessary for reduction to occur. Therefore, the species oxidized is called the reducing agent. The species reduced is called the oxidizing agent. Oxidation cannot occur without the corresponding reduction process and vice versa.

Let's look at an example. Fe^{2+} reacts spontaneously with Ce^{4+} according to the equation,

$$Fe^{2+} + Ce^{4+} \rightarrow Fe^{3+} + Ce^{3+}$$

The oxidation number of iron changes from +2 to +3, an increase in the positive value of the oxidation number. Oxidation numbers will increase in positive value as the chemical species gives up or loses electrons. We recognize that an element in a chemical equation is oxidized whenever that element's oxidation number increases in positive value in the course of the reaction (from reactant to product side of the equation.)

The oxidation number for cerium, however, undergoes a decrease in positive value in this reaction. Reduction is seen as the decrease in the positive value of the oxidation numbers.

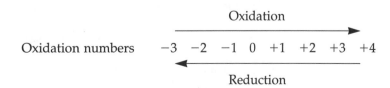

The overall redox reaction can be broken down into the two half reactions, one for oxidation and one for reduction:

$$Fe^{2+} \rightarrow Fe^{3+} + e^- \qquad \text{oxidation}$$

$$Ce^{3+} + e^- \rightarrow Ce^{4+} \qquad \text{reduction}$$

In these half reactions, we see that Fe^{2+} loses one electron and Ce^{3+} gains one electron. In the overall reaction, the number of electrons lost must equal the number of electrons gained. In this case, the requirement is satisfied, so the half reaction equations can be added to give the overall equation:

$$Fe^{2+} + Ce^{4+} \rightarrow Fe^{3+} + Ce^{3+}$$

Electrical Potential and Redox Reactions

Metals have a strong tendency to give up electrons and form positive ions, but the strength of this electron-pushing tendency varies from metal to metal. Thus, metals can be ranked according to their ability to give up electrons (i.e., according to their oxidation potential). This ranking is called the activity series for metals. The activity series is often used to predict whether a particular reaction is likely to occur. As part of this experiment, you will measure the oxidation potentials for the metals selected by your instructor and generate your own mini–activity series.

Many texts include tables of standard reduction potentials. The word "standard" refers to measurements made for pure solids, solutions of 1 M concentration, and gases at 1 atm pressure. A standard measurement is usually indicated by a superscript "0," so a standard reduction potential is written $\mathscr{E}°$. These tables list reduction half reactions according to their tendency to occur.

The standard reduction potential ($\mathscr{E}°$) for each half-reaction is measured against the potential for hydrogen to be reduced, assuming the hydrogen potential to be zero. Any chemical species that is reduced more easily than hydrogen is assigned a positive value. Any chemical species that is reduced less easily than hydrogen is assigned a negative value.

When two half-reactions are put together, the one with the higher positive reduction potential will in fact be reduced. The second half-reaction then must be oxidized. When we combine two half-reactions to form an overall equation for a spontaneous reaction, we expect that the overall potential $\Delta\mathscr{E}°$ will be positive. To obtain the equation for the spontaneous reaction, we substract the half-reaction with a lower $\mathscr{E}°$ from the reaction with a higher $\mathscr{E}°$. Similarly, we can get the reaction's overall potential by subtracting the value of the lower $\mathscr{E}°$ from the value of the higher $\mathscr{E}°$.

Let's look at an example. Suppose a galvanic cell contains $Ag^+|Ag$ as one half-cell and $Sn^{2+}|Sn$ as the other. The standard half-reactions and the standard reduction potentials are

$$Ag^+ + e^- \rightarrow Ag \qquad \mathscr{E}° = 0.7996 \text{ V}$$

$$Sn^{2+} + 2e^- \rightarrow Sn \qquad \mathscr{E}° = -0.1364 \text{ V}$$

A comparison of the $\mathscr{E}°$ values clearly indicates that Ag^+ has a greater tendency to be reduced than Sn^{2+}. Ag^+ is reduced to Ag at the cathode. Sn is oxidized to Sn^{2+} at the anode. The spontaneous reaction that occurs is

$$2\, Ag^+\, (aq) + Sn\, (s) \rightarrow 2\, Ag\, (s) + Sn^{2+}\, (aq) \qquad \textbf{(1)}$$

and the measured cell potential difference is, $\Delta\mathscr{E}°_{cell} = \mathscr{E}°_{Ag} - \mathscr{E}°_{Sn} = 0.7996 - (-0.1364) = 0.9360$ V.

Electrochemical Cells

A simple galvanic cell (Figure G-1) consists of two beakers, each with a metal electrode, appropriate wiring, and a voltmeter. The beakers contain aqueous solutions. If a half-reaction involves a pure element, then the electrode is made of that element. Otherwise, an inert electrode is used. The electrode in the beaker where reduction occurs is called the cathode and the electrode in the beaker where oxidation occurs is the anode.

Each electrode is connected to the voltmeter by alligator clips and metal wiring. The voltmeter measures the voltage generated by the redox reaction. The voltage reading will be positive when the electrodes are connected properly for a spontaneous reaction. A spontaneous redox reaction occurs when the species with the higher reduction potential is connected as the cathode. Otherwise, the voltage reading will be negative. The meter reading will be positive when the cathode is connected to the $(+)$ outlet and the anode is connected to the $(-)$ outlet. Physically, a negative voltage reading means you have connected the wrong electrode as cathode. This is equivalent to reversing equation (1). When equation (1) is reversed, the measured cell potential difference becomes, $\Delta\mathscr{E}°_{cell} = \mathscr{E}°_{Sn} - \mathscr{E}°_{Ag} = -0.1364 - 0.7996 = -0.9360\text{ V}$. The absolute value of $\Delta\mathscr{E}°_{cell}$ is the same in both cases, but the sign is different. The sign of $\Delta\mathscr{E}°_{cell}$ is positive for a spontaneous reaction and negative for a nonspontaneous reaction. An electrochemical cell becomes a galvanic cell when the electrodes are connected so that the spontaneous oxidation–reduction reaction always occurs. An electrolysis cell uses an electric current to force an otherwise nonspontaneous reaction to occur.

A typical electrochemical cell is shown in Figure G-1. The direction of electron flow from the anode to the cathode is also shown. The function of the salt bridge is to maintain electroneutrality in the system as the electrons are transferred from the anode to the cathode during the course of the reaction. The salt bridge contains ions that do not participate in the redox reaction but do migrate in response to the electron flow. As electrons move away from the anode, positive charge builds up around

FIGURE G-1 SCHEMATIC OF A GALVANIC CELL

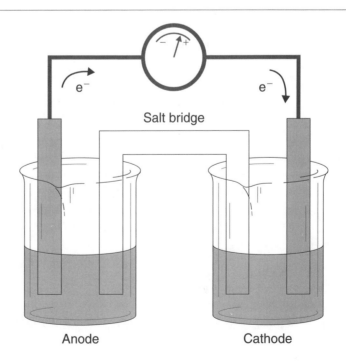

Salt bridge

Anode   Cathode

it; as electrons move toward the cathode, it becomes negatively charged. This causes negative ions from the salt bridge to flow toward the anode while positive salt-bridge ions flow toward the cathode.

Galvanic cells similar to that shown in Figure G-1 can easily be assembled and used. However, building such cells with beakers requires a large amount of solution. Your instructor may have alternate materials that use much smaller volumes.

In today's experiment, you will measure the potential difference between various half reactions. This will give you experimental values that you can use to rank your metals according to their oxidation potential. Your instructor will tell you what kind of materials to use in the construction of your electrochemical cell.

The Nernst Equation

For measurements taken under standard conditions (1 atm, 1 M solutions), $\Delta \mathcal{E}°_{cell}$ measures the electric potential difference between the half cells. For measurements taken under non-standard conditions (the usual laboratory situation), the Nernst equation is used to calculate $\Delta \mathcal{E}_{cell}$. The Nernst equation gives us the relationship between the overall cell potential difference for a redox reaction ($\Delta \mathcal{E}_{cell}$) and the concentrations of the metal-ion solutions.

The Nernst equation is $\Delta \mathcal{E}_{cell} = \Delta \mathcal{E}°_{cell} - ``(RT)/(n\mathcal{F})"\ln Q$, where $\mathcal{F}$ is the Faraday constant, R is the gas constant, n is the number of electrons transferred, and Q is the reaction quotient. Sometimes the natural log (ln) is converted to the base ten log by $\ln x = 2.303 \log x$. and the constants $(RT/\mathcal{F})$ are evaluated using $R = 8.315$ J K^{-1} mol^{-1}, $T = 298.15$ K, and $\mathcal{F} = 96{,}485$ C mol^{-1}. Then the equation becomes

$$\Delta \mathcal{E}_{cell} = \Delta \mathcal{E}°_{cell} - \left(\frac{0.0592 \text{ V}}{n}\right) \log Q$$

In this form, galvanic cells are used to determine the concentrations of the metal ions present under conditions other than standard conditions.

Suppose we again examine equation (1). We have already considered the standard half-cell potentials given by

$$Ag^+ (aq) + e^- \rightarrow Ag (s) \qquad \mathcal{E}° = 0.7996 \text{ V}$$

$$Sn^{2+} + 2e^- \rightarrow Sn \qquad \mathcal{E}° = -0.1364 \text{ V}$$

and the spontaneous reaction that occurs.

$$2 \text{ Ag}^+ (aq) + Sn (s) \rightarrow 2 \text{ Ag} (s) + Sn^{2+} (aq) \qquad \Delta \mathcal{E}°_{cell} = 0.9360 \text{ V}$$

For this reaction, the Nernst equation is

$$\Delta \mathcal{E}_{cell} = \Delta \mathcal{E}°_{cell} - \left(\frac{0.0592 \text{ V}}{n}\right) \log \frac{[Sn^{2+}]}{[Ag^+]^2}$$

Now, if we wish to determine the concentration of Ag^+ when $\Delta \mathcal{E}_{cell} = 0.9408$ and $[Sn^{+2}] = 0.010$ M, then $[Ag^+]$ is easily obtained by solving the equation,

$$0.9408 = 0.9360\text{V} - \frac{0.05916}{2} \log \frac{0.010}{[Ag^+]^2}$$

$$-0.16227 = \log 0.01 - 2 \log [Ag^+]$$

Solving, we get

$$\log [Ag^+] = -0.9189$$

$$[Ag^+] = 0.121 \text{ M}$$

Construction of an Electrode

In the second part of this experiment, you will build your own electrode. Electrodes are the vehicles by which electrons are shuttled from oxidized to reduced species in a redox reaction. It is possible to make a simple electrode assembly in such a way that it contains the entire electrochemical cell. This kind of electrode is called a combination electrode. One half-cell serves as a reference electrode and consists of a metal in contact with a solution that contains a known concentration of the same metal ion. The electrode you will make will use a copper wire immersed in a solution of copper(II) sulfate solution that has a concentration that is both known and constant. This reference electrode provides a constant potential for the cell. The special assembly, shown in Figure G-2, also has a second metal electrode, in this case, silver. The silver electrode is called the working electrode because it does the "work" in measuring the concentration of an unknown solution that contains silver ion.

When the $Ag^+|Ag$ half-cell is compared to a reference half-cell, the overall cell potential is given by $\Delta\mathscr{E}_{cell} = \mathscr{E}_{Ag} - \mathscr{E}_{ref}$. You should note that, although the concentration of the copper solution is held constant, it may vary from electrode to electrode so $\mathscr{E}_{ref}$ is different for each electrode. For a given electrode, however, the potential of the reference electrode is *constant*. This allows us to focus on only the portion of the Nernst equation for the working electrode—in this case, the silver

FIGURE G-2   A SELF-CONSTRUCTED ELECTRODE

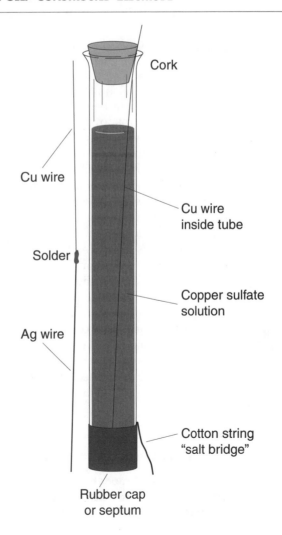

electrode. The working electrode is always the cathode; the reference electrode is the anode.

Using the Nernst equation, we can evaluate the $\mathscr{E}°_{cell}$ for Ag and the relationship simplifies to

$$\Delta\mathscr{E}_{cell} = \Delta\mathscr{E}°_{cell} + \frac{0.05916}{n} \log[Ag^+]$$

where n is 1, the moles of electrons transferred by silver. This equation is of the form $y = mx + b$. The concentration of silver, $[Ag^+]$, can be determined as a function of $\Delta\mathscr{E}_{cell}$ by measuring $\Delta\mathscr{E}_{cell}$ for various standard solutions of Ag^+ and generating a calibration curve. A linear plot results when $\Delta\mathscr{E}_{cell}$ (which corresponds to the y in the general equation) is graphed against $+ \log[Ag^+]$ (which stands in the place of the x in the general equation

$$\Delta\mathscr{E}_{cell} = \Delta\mathscr{E}°_{cell} + \frac{0.05916}{n} \log[Ag^+]$$

$$y \quad = \quad b \quad + \quad (m) \quad\quad x$$

The slope of the line is $0.05916/n$, and the y-intercept is $\Delta\mathscr{E}°_{cell}$. By measuring $\Delta\mathscr{E}_{cell}$ for various known concentrations of silver, a calibration curve can be produced and used to determine $[Ag^+]$ for a solution of Ag^+ of unknown concentration.

Because the copper reference potential may vary from electrode to electrode, you must create a calibration curve for each new electrode you make and use.

Calibration Curves

The equation, $\Delta\mathscr{E}_{cell} = \Delta\mathscr{E}°_{cell} + (0.05916/n) \log[Ag^+]$, shows that there is a linear response between $\Delta\mathscr{E}°_{cell}$ and $\log[Ag^+]$. The voltmeter that produces the measurement of $\Delta\mathscr{E}°_{cell}$, then, can be calibrated and used to measure $\log[Ag^+]$ for solutions in which $[Ag^+]$ is unknown. The voltmeter is calibrated by measuring $\Delta\mathscr{E}°_{cell}$ for several solutions with known concentrations of Ag^+. This generates several pairs of linked $\Delta\mathscr{E}°_{cell}$ and $[Ag^+]$ measurements. If we find $\log[Ag^+]$, then we have x, y data in the form of $\log[Ag^+], \Delta\mathscr{E}_{cell}$. With several x, y data points, we can generate a graph of the data. The curve connecting the data points in the graph represents the response of the voltmeter to various known concentrations of Ag^+. Because of the form of this equation, we expect a linear response. To use the calibration curve, find the y-intercept and the slope of the line. Substitute these values into the equation for $\Delta\mathscr{E}_{cell}$. This transforms what was a general equation for this type of reaction to a specific equation for your system. To determine an unknown Ag^+ concentration, then, simply measure $\Delta\mathscr{E}°_{cell}$ for the unknown solution, substitute this value into the equation from your calibration curve and solve for $[Ag^+]$.

Caution: Observe proper caution in making the electrical connections. In particular, never touch bare wiring that is part of a completed electrical circuit. Many solutions containing metal ions, especially Pb and Ag, are hazardous and must be handled with care. All solutions should be handled and disposed of properly.

Procedure

Part I: Construction of a Galvanic Cell (Group Work)

Several different models of voltmeters may be used for this experiment. Check with your laboratory instructor to be sure that you have the correct model and that it is connected properly to obtain direct current measurements. You must use a direct current (DC) voltmeter and set it to measure volts. To set up the meter for voltage measurements, you will also need two electrical leads with jacks at one end and alligator clips at the other.

This part of the procedure should be done in pairs. Common metals you may use are silver, copper, iron, lead and zinc. Solutions of silver ion, copper(II) ion, iron(II) ion, zinc ion and lead(II) ion will be available.

A number of half-cells can be set up by using beakers (see Figure G-1) or smaller cells. Strips of filter paper may be used as salt bridges connecting the two half-cells.

Part II: Measuring $\Delta\mathscr{E}_{cell}$ (Group Work)

Be certain that the alligator clips are clean before connecting to the electrodes. You can clean them by pulling some sandpaper between their jaws. When you are confident that all metals are clean and properly placed, connect the alligator clips to any two electrodes. Remember that a galvanic cell is one in which the redox reaction proceeds spontaneously and that a spontaneous redox reaction has $\Delta\mathscr{E} > 0$.

Begin at the highest DC-V setting and gradually lower the setting until you get a reading. If your reading is negative, you have the leads connected to the wrong outlets. Exchange them. The meter reading will be positive when the cathode is connected to the (+) outlet and the anode to the (−) outlet of the meter. *Record the voltage and label* which of the two metal electrodes is the cathode for this pairing. Disconnect the alligator clips, *clean them again*, and record a second measurement. If these two measurements don't agree within 5%, repeat the cleaning and measurements again until you obtain reasonable agreement between measurements. Repeat this procedure for all possible pairings of half-cells.

Cleanup. Clean and return all metals, meter, leads, and weighing dish. Do *not* flush any solutions down the drain. Dispose of them properly.

Part III: The Electrochemical Measurement of [Cu²⁺] (Individual Work)

Each student should assemble an electrochemical cell composed of standard Cu^{2+}/Cu and Zn^{2+}/Zn and measure the cell potential difference, $\Delta\mathscr{E}°_{cell}$.

Obtain strips of copper and zinc metals. Clean them using the same procedure as before. Also obtain an appropriate volume of zinc and copper ion solutions. Immerse the zinc metal into the zinc solution and the copper metal into the copper solution. Devise a salt bridge, using filter paper or other available materials. Check with your instructor on this. Attach the alligator clips so that the cell potential is positive. Allow a few minutes for the system to reach equilibrium and then record the cell potential difference. Obtain a sample of unknown Cu^{+2} solution and repeat the procedure.

Part IV: Constructing a Working Electrode (Group Work)

This part of the procedure is best done in groups of two. The assembly pictured in Figure G-2 contains two electrodes: a $Cu^{2+}|Cu$ reference half-cell and a silver electrode. An electric potential difference will develop between the two electrodes when the silver electrode is put in contact with a solution containing Ag^+, completing the circuit and forming the electrochemical cell.

To construct the electrode assembly,

- Obtain enough materials to make two electrodes.
- Obtain the copper and silver wire electrodes and assemble. Each assembly should contain two electrodes: one copper wire and one copper–silver wire (soldered together).

- Insert the copper wire into a length of glass tubing and secure with a stopper.
- Fasten the silver–copper soldered wire to the outside of the tubing with tape.

Fill the reference electrode with 0.1 M $CuSO_4$ solution, insert thread, and finally stopper with the septum.

When using this electrode assembly, it is essential that the string, which functions as the electrochemical salt bridge, make contact with the solution inside and outside of the reference electrode.

Part V: Preparation of the Calibration Curve (Group Work)

Obtain about 25 mL of a standardized solution of Ag^+. Prepare four dilutions such that you end up with four 10-mL solutions, each with a different concentration. The solutions should have concentrations of between 0.12 M and 0.05 M. Each person in the group should prepare at least one of the solutions by dilution from the stock solution. Do the calculations for these dilutions now.

Prepare a table in which to record the concentration of each solution after dilution and the $\Delta\mathscr{E}°_{cell}$ measured with your electrode. Obtain a voltmeter and two leads. Make the appropriate connections. Connect leads so that the voltage reading is positive.

Measure $\Delta\mathscr{E}°_{cell}$ for each of the solutions.

Part VI: Measurement of an Unknown Concentration of Ag^+ (Individual Work)

Obtain a solution of unknown silver concentration. Record its identifying number or letter. Measure the cell potential difference following the above procedure. Repeat the measurement several times until you are satisfied that your voltage reading is constant. Each member of the group should measure a different unknown solution.

REPORT

Galvanic Cell Experiments (Individual Work)

1. Write a balanced equation for all ten redox reactions you measured. Circle the metal that served as the cathode in each reaction.
2. Calculate the theoretical value of $\Delta\mathscr{E}_{cell}$ for each reaction by using the Nernst equation, a table of standard reduction potentials, and the actual concentrations of the solutions used.
3. Compare these theoretical cell potential differences with those you *actually* measured and calculate the percent error for each cell. Comment on possible sources of error in your measurements.
4. Answer the following questions:
   a. For one pairing that initially gave a negative voltage reading, write the redox equation, identify the anode and cathode, and give the direction of flow of electricity in the external circuit.
   b. What is the purpose of the KNO_3 salt bridge?
   c. A galvanic cell cannot generate electricity forever. List two chemical reasons you can think of for why a galvanic cell may go "dead."

Galvanic Cell Experiments (Group Work)

Discuss with your group the best way to determine the ordering of the five metals, from the one that makes the best cathode to the one that is the worst. Include this listing in your lab report.

Unknown Concentration of Cu²⁺

1. Use standard reduction potentials for $Zn^{2+} \mid Zn$ and $Cu^{2+} \mid Cu$ to calculate $\Delta\mathscr{E}°_{cell}$.

2. Use the Nernst equation and your experimental results to calculate the concentration of the copper(II) ion solution.

3. Obtain the actual value for $[Cu^{2+}]$ from your instructor. How does your experimental value compare to the actual value?

4. Critique your experimental setup. What could have been improved?

5. Critique your personal lab techniques. What could you have done better?

Construction of an Electrode

1. For your final report, include a sketch of your electrode assembly.

2. Briefly list and discuss any problems you encountered with the procedure.

3. Prepare a graph of $\Delta\mathscr{E}_{cell}$ versus $\log[Ag^{+}]$. Determine the y-intercept and the slope of the line graphed. Calculate the silver concentration of your unknown solution from its value of $\Delta\mathscr{E}_{cell}$ using the calibration curve equation.

4. Compare the theoretical values for $\Delta\mathscr{E}°_{cell}$ and $0.05916/n$ with the values obtained for the y-intercept and slope from your calibration curve.

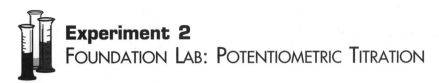

Experiment 2
FOUNDATION LAB: POTENTIOMETRIC TITRATION

Pre-Laboratory Assignment   **Due Before Lab Begins**

NAME:_____

Complete these exercises after reading the experiment but before coming to the laboratory to do it.

1. Look up the half-cell reaction for a silver–silver chloride electrode. You can find this together with its $\mathscr{E}°$ value in a table of standard reduction potentials.

2. A 50.0-g sample is found to contain 3.48×10^{-3} g of Cl^-. Express this in concentration units of parts per million (ppm) by mass.

3. Describe what happens when silver nitrate comes into contact with your skin.

4. The following is some data for the potentiometric titration of chloride with 0.0500 M $AgNO_3$. It is important to keep track of the changes, $\delta(\Delta\mathscr{E})/\delta V$, as the titration progresses.

a. Complete the table by calculating the remaining $\delta(\Delta\mathscr{E})/\delta V$ ratios.

Volume $AgNO_3$, mL	$\Delta\mathscr{E}_{cell}$, V	$\delta(\Delta\mathscr{E})/\delta V$*
5.0	0.396	none
15.0	0.385	$(0.385 - 0.396)/(15.0 - 5.0) = -0.0011$
20.0	0.373	
22.0	0.358	
23.0	0.351	
23.5	0.340	
23.7	0.316	
24.0	0.233	
24.10	0.194	
24.20	0.183	
24.30	0.170	
24.40	0.158	
24.50	0.144	
24.60	0.135	
24.70	0.120	
25.00	0.107	
25.50	0.085	
26.00	0.062	$(0.062 - 0.085)/(26.00 - 25.50) = -0.046$

* Do not be confused by the notation employed here. Since $\Delta\mathscr{E}$ is already a potential difference and you are measuring changes in $\Delta\mathscr{E}$ by subtracting one potential difference from another, we denote this change in $\Delta\mathscr{E}$ as $\delta(\Delta\mathscr{E})$. We also denote the corresponding change in volume as δV.

b. The largest $\delta(\Delta\mathscr{E})/\delta V$ ratio indicates the endpoint of the reaction. Between which two volumes does this occur? The titrant volume at the endpoint lies halfway between these two volumes. What is the titrant volume at the endpoint?

c. Compute the number of moles of Ag^+ in the titrant volume given in part b.

d. Write the balanced equation that occurs between Ag^+ (aq) and Cl^- (aq). How many moles of Cl^- are in the unknown solution?

Experiment 2
FOUNDATION LAB: POTENTIOMETRIC TITRATION

BACKGROUND

A titration monitors a reaction by tracking a measurable difference in some property. You may be familiar with acid-base titrations. Acid–base titration is only one of several types of titration that chemical professionals use in chemical analysis. Titration curves are similar in appearance but vary according to the signal that is being monitored. In each titration curve, a titrant is added to the system of interest while a particular property is monitored. The researcher records both the volume of titrant added and the signal response of the instrument used to monitor the titration. For an acid–base titration, the signal is either pH or an indicator that changes color at the equivalence point of the titration.

For a potentiometric titration, the signal monitored is the electrochemical cell potential, $\Delta\mathscr{E}_{cell}$. The equivalence point of the titration marks the point at which the chemical species being titrated is completely reacted. This point is judged to be the center of the steeply rising portion of the titration curve.

In this experiment, you will measure the potential difference that exists in an electrochemical cell as a function of titrant volume—this is called a potentiometric titration. You will use the electrode assembly you just constructed as the indicator electrode, and you will generate a titration curve by plotting the cell potential difference in volts versus the volume of titrant added. There are several ways to determine the endpoint of the reaction from the titration curve, but the most common method is to determine visually the inflection point of the curve. The inflection point is the point at which the curve changes direction or sign (the curvature goes from being concave to convex or vice versa)—roughly in the middle of the steeply rising portion of a titration curve.

For many systems—such as those with deep color, turbidity, or unsuspected, secondary ions, a potentiometric titration gives more accurate results than do chemical indicators. An obvious disadvantage of potentiometric titration is the greater length of time required to do the entire titration.

In this experiment, you will measure the chloride concentration in an aqueous sample by titrating the chloride sample with a silver nitrate solution of known concentration. The reaction, then, is

$$Ag^+ \ (aq) + Cl^- \ (aq) \rightarrow AgCl \ (s)$$

You can estimate the volume of silver nitrate to be added by evaluating the ratio of $\delta(\Delta\mathscr{E})$ to δV after each addition. At the beginning of the titration, $\delta(\Delta\mathscr{E})/\delta V$ is small. $\Delta\mathscr{E}$ changes slowly because the added Ag^+ is precipitated as AgCl. However, there is a larger change in $\Delta\mathscr{E}$ at the endpoint, as the ratio becomes much larger because the Ag^+ is no longer precipitated. At the equivalence point, $\delta(\Delta\mathscr{E})/\delta V$ is at a maximum. After the endpoint the ratio again becomes small. You will obtain a better endpoint if you keep track of $\delta(\Delta\mathscr{E})/\delta V$ and decrease the volumes added as $\delta(\Delta\mathscr{E})/\delta V$ increases in magnitude.

The concentration of chloride can be calculated from the titrant volume at the endpoint and from the stoichiometry of the reaction.

Caution: Silver nitrate is hazardous. Handle it carefully. Immediately rinse off any silver nitrate that comes into contact with your skin. Dispose of used solutions according to usual laboratory procedures.

PROCEDURE *

Part I: Constructing a Working Electrode (Group Work)

Use the electrode you constructed in the first experiment or construct a new one. See directions in experiment 1 of this module. This part of the procedure is best done in groups of two.

Part II: The Potentiometric Titration (Group Work)

This part of the experiment is best accomplished by a group of two. First you will practice the potentiometric titration with solutions of known concentrations. Then you will repeat the titration to determine an unknown concentration of Cl^-.

Obtain and clean a buret for the silver nitrate solution. Rinse it with the stock solution of silver nitrate, and secure in place for the titration. Position the buret so that splashing is minimized when the silver nitrate solution (the titrant) is added.

Put 50.00 mL of 0.0010 M chloride solution (accurate to ±0.05 mL) into a clean beaker.

> **NOTE:** The beaker must be large enough to accommodate the tip of the buret containing the titrant, the electrode assembly, and a magnetic stirring bar.

Secure the electrode assembly in place and immerse into the sample solution. Make the necessary electrode connections to the voltmeter, begin magnetic stirring, and record both the initial volume and $\Delta\mathscr{E}_{cell}$ reading. Add a small volume (2–3 mL) of silver nitrate solution to the beaker containing the chloride solution. Record the buret volume after this addition. Record the voltage when it has stabilized to within a fluctuation of several millivolts. This may require a wait of 30 to 60 seconds or more.

 Caution: If your electrode has the silver wire soldered to the copper wire, be very careful not to let the solution in the beaker ever touch the solder. Erratic fluctuations of voltage occur when the solder becomes immersed in the solution. The soldered point must remain above the solution at all times.

Continue adding the titrant, recording the volume added and the new voltage reading each time. Calculate $\delta(\Delta\mathscr{E})/\delta V$ after each addition. Decrease titrant volumes to 0.1 mL as the endpoint is approached. (Remember that this ratio becomes larger near the endpoint.) Continue the titration about 2 or 3 mL beyond the endpoint.

Part III: Determination of Chloride in a Solution of Unknown Concentration (Individual Work)

Each group member will take a turn using the electrode assembly to record the titration curve of a chloride solution with an unknown concentration. While one partner does the actual titration, the second partner should calculate $\delta(\Delta\mathscr{E})/\delta V$ and provide direction for the titration. Increments of titrant should be decreased when the $\delta(\Delta\mathscr{E})/\delta V$ becomes large and should be increased when $\delta(\Delta\mathscr{E})/\delta V$ becomes small. Obtain about 75.00 mL of your assigned unknown. Use 50 mL of the unknown solution in the titration.

---

* This procedure is modeled on one reported by George Lisensky and Kelly Reynolds; *J. Chem. Ed.*, **68**, *4*, **1999**, 334–335.

When you are finished, disassemble the electrode assembly, clean the parts, and return them to the supply area.

REPORT

1. Prepare a graph of your data, plotting voltage on the vertical axis and milliliters of titrant added on the horizontal axis.

2. Determine the endpoint of your titration curve visually. What volume of silver nitrate had been added when the endpoint was reached? How many moles of Ag^+ are contained in this volume?

3. Write a balanced equation for the reaction between silver ions and chloride ions, and use it to calculate the moles of chloride in your unknown. With dilute aqueous solutions such as your sample of chloride ion, we can assume that the solution density is the same as that of pure water. Assuming a density of 1.00 g/mL for your chloride solution, express your chloride concentration in terms of parts per million (ppm) by mass. Include this concentration in your final report.

4. The endpoint of the titration can be graphically determined in a second manner. Prepare a graph of $\delta(\Delta \mathscr{E})/\delta V$ values vs. volume of titrant added. Estimate the endpoint as the volume at the maximum value of $\delta \Delta \mathscr{E}/\delta V$.

Experiment 3
APPLICATION LAB: THE GLUCOSE BIOSENSOR

Pre-Laboratory Assignment   **Due Before Lab Begins**

NAME:_____

1. a. When the blood glucose level ($C_6H_{12}O_6$) is too low (less than 50 mg/dL) or
       too high (greater than 300 mg/dL), patients are urged to consult their
       doctor. Convert 50 mg/dL and 300 mg/dL to units of mol/L.

   b. Given a 0.500 M stock solution of glucose, how would you prepare 100 mL
       glucose solutions with concentrations of 85 mg/dL, 165 mg/dL, and 325
       mg/dL?

2. Write the balanced equation for the oxidation of $[Fe(CN)_6]^{4-}$ to $[Fe(CN)_6]^{3-}$.
   Determine the oxidation state of Fe in both anions. What is the standard
   reduction potential for this redox couple?

3. a. Suppose you obtain the calibration curve in Figure G-3 for your experiment.
       Which of the data points can not be used to obtain the slope of the line
       generated? Circle these.

FIGURE G-3

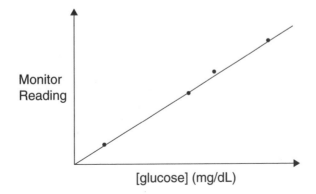

b. It is often better to use an equation rather than to read the graph itself to interpret calibration data. How can you obtain an equation for this experiment from this calibration graph? (*Hint:* Refer to the previous experiment for a discussion of equations and graphs.)

4. One of the reactions that occurs on a glucose test strip is

glucose + enzyme(oxidized form) → gluconic acid + enzyme (reduced form)

In this equation, the oxidized and reduced form of the enzyme are labeled. Label the remaining formulas in this equation as the oxidized and reduced form of glucose.

Experiment 3
APPLICATION LAB: THE GLUCOSE BIOSENSOR

BACKGROUND

Over the past forty years, over-the-counter biosensors have drastically increased in importance as they have become cheaper and more reliable for consumer use. A biosensor is an instrument that can measure (or sense) some biochemical substance by tracking or monitoring a signal such as change in pH or in current. A glucose biosensor uses miniaturized microelectrodes to monitor blood glucose concentration as a function of current. A previous experiment in this Group depended on spontaneous reactions that produced a cell potential difference. They were both potentiometric experiments. The glucose biosensor depends on current produced in an electrolysis cell. As current is measured in units of amperes (coulombs per second), it is called an amperometric experiment.

The glucose testing kit has two main components, the monitor (instrument) that measures blood glucose level and test strips onto which the patient places a drop of blood for testing in the monitor. The test strips contain all of the necessary components of the cell except the reactant glucose. Two thin microelectrodes are positioned parallel to one another and run the length of the test strip. The end of the test strip that is inserted into the monitor has the ends of these electrodes exposed so they are easily seen by the naked eye. On the opposite end of the test strip is a small yellow square "testing spot." The testing spot is in direct contact with the microelectrodes. A typical test strip* has three very thin layers of materials applied to the testing spot in a process similar to silk screening. The outer layer is a polycarbonate film that is permeable to glucose but not permeable to larger proteins or blood contaminants. The middle layer contains an enzyme and other chemicals necessary for the reaction. The inner layer is made of cellulose acetate, which is permeable to smaller molecules. This last layer is important to systems that monitor production of a side-product such as H_2O_2.

Many over-the-counter glucose biosensors use a glucose catalase enzyme, but this enzyme needs a supply of oxygen to function properly.[†] Oxygen dependence has been a serious problem with glucose test kits that use this enzyme because insufficient oxygen results in erroneous glucose concentration readings. Many pharmaceutical companies have investigated various ways of "building in" excess oxygen supplies by adapting the chemicals and/or the type of electrodes used on the test strip. Other manufacturers have replaced the enzyme with one that does not have this oxygen requirement. The test strips in your glucose kits use an enzyme called glucose dehydrogenase, which has no oxygen dependence.

The glucose biosensor employs an electrochemical method that measures the current that flows when a battery applies a potential to a microelectrode. The resulting current has a linear dependence on the glucose concentration. When the test strip (with a drop of blood or, in our case, a glucose sample) is inserted into the monitor, the voltage is applied to the microelectrodes. A series of oxidation–reduction reac-

---

* This description is adapted from material provided by Boehringer Mannheim Corporation and applies to the Advantage test strips that are used with Accu-Chek Advantage monitor. Details are expected to be similar for other glucose test strips.

[†] A good discussion of the oxygen dependence of glucose biosensors is given in *Chemical & Engineering News*, Feb. 23, 1998; 61–64.

tions that occur at this time generate a current. The current has a direct relationship to the glucose concentration in the original sample.

The Electrolysis Reaction

The reactions that occur on the test strip rely on an enzyme that is selective for glucose and on the anion, $[Fe(CN)_6]^{3-}$. The $[Fe(CN)_6]^{3-}$ anion acts as a mediator to enhance the oxidation reduction reaction at the electrode. A mediator is a redox couple that has very fast kinetics. It speeds up or enhances the exchange of electrons between the electrode and the enzyme. For our system, the mediator is the redox couple, $[Fe(CN)_6]^{3-}$–$[Fe(CN)_6]^{4-}$. $[Fe(CN)_6]^{3-}$ is called hexacyanoferrate(III); it is the oxidized form of the ion. $[Fe(CN)_6]^{4-}$ is called hexacyanoferrate(II); it is the reduced form of the ion.

The reaction sequence consists of three steps: an enzymatic reaction, a mediator reaction, and an amperometric reaction. The following equations represent a simplification of the reaction.

- *Enzymatic reaction*: Glucose from the sample diffuses through the layers towards the electrode when a sample is placed on the test strip. In this step, the enzyme oxidizes the glucose:

  glucose + enzyme(oxidized form) → gluconic acid + enzyme (reduced form)

- *Mediator reaction*: In this step, the active form of the enzyme is regenerated by oxidation by $[Fe(CN)_6]^{3-}$.

  enzyme (reduced form) + $[Fe(CN)_6]^{3-}$ →

  $$[Fe(CN)_6]^{4-} + \text{enzyme (oxidized form)}$$

- *Amperometric reaction*: The glucose sensor is technically an enzyme-based biosensor. A voltage or potential (generally 300 mV) is applied to the micro-electrodes and a current is produced by the oxidation of $[Fe(CN)_6]^{4-}$:

  $$[Fe(CN)_6]^{4-} \rightarrow [Fe(CN)_6]^{3-} + e^-$$

You will notice that these reactions are cyclic in nature and relate directly back to the original glucose. To see this better, look at the sequence in reverse. In the amperometric reaction, the $[Fe(CN)_6]^{3-}$ produced goes on to regenerate the active form of the enzyme, which in turn reacts with the glucose. Both the enzyme and the $[Fe(CN)_6]^{3-}$, sometimes called a cofactor for the enzyme, are regenerated and we obtain a quantitative measure of the glucose originally present.

The Glucose Test Kit

The test kit (shown in Figure G-4) is comprised of a monitor and test strips. Each box of test strips contains a coding strip that matches the code on the test strips. Before using test strips from a new box, the coding strip for that box must be inserted into the monitor so that the monitor will correctly interpret test results. Your lab instructor will direct you to do this coding, if necessary. The top of the monitor has a display screen, an on/off button, a memory button, and an open space for test strip insertion.

Follow the directions accompanying your monitor kit for test strip insertion and placement of sample. In general, the test strip is inserted into the monitor first. Only the electrode portion of the test strip will be inside the monitor when the test strip is inserted properly. The sample area of the test strip remains outside of the moni-

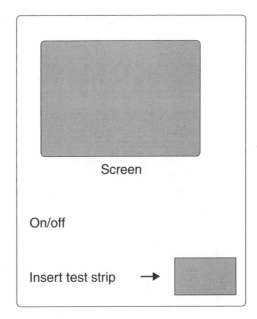

Screen

On/off

Insert test strip   →

FIGURE G-4

tor. With the test strip inserted, place a drop of sample on the test spot. The glucose concentration will show on the screen after a short time. Remember all the processes that must occur before a current is generated.

Range of Results

The normal blood glucose range for a fasting, nondiabetic adult is 70–105 mg/dL, whereas normal blood glucose levels 1 to 2 h after eating may be 110–180 mg/dL. The following ranges are suggested when using the glucose monitor:

45–75 mg/dL	*low blood glucose, seek help or treatment*
110–180 mg/dL	normal range
255–345 mg/dL	*high blood glucose, seek help or treatment*

For the greatest reliability, the sample must completely cover the test spot.

 Caution: No blood or blood products are used in this experiment, only commercially available solutions containing glucose. Exercise proper caution when making connections to the DC source for the electrolysis cell. Never touch any part of the metal prongs on the plug when inserting the plug into an electrical outlet.

PROCEDURE

Part I: Formation of Groups

This procedure is best done in groups of two. Both individuals are responsible for setting up the equipment needed in this experiment and for measuring some data. In most cases the data measured by each member of the team involve different solution concentrations and/or amounts. Throughout the experiment (or at the end of the experiment) you should compare your results with those of your team member. This serves as a check of the reproducibility of your data and as an aid in reaching a conclusion.

Part II: Investigation of an Electrolysis Reaction*

A. Setting Up the Electrolysis Cell (individual work)

Obtain two copper electrodes from your lab instructor or from the supply area of your lab. Clean them thoroughly first with sandpaper or steel wool and then by immersion in dilute HCl for 30 s. Finally, rinse both electrodes with deionized water, dry with a small amount of acetone, and carefully set aside.

Each team member should obtain two beakers. One team member should obtain a 300-mL beaker and a 150-mL beaker. The second team member should obtain a 200-mL beaker and a 100-mL beaker. Between them, the team members should have four different sizes of beakers.

Bend the copper electrodes so that they fit inside the smaller beaker, almost touching the bottom. The bent part of the copper should fit over the side of the beaker so that the electrode remains stationary. Scratch a mark on both electrodes at an equal distance from the bottom of the beaker to serve as a fill-line for the solution when it is added.

Remove the electrodes from the smaller beaker and secure in place in the larger beaker. Add acidified copper(II) sulfate stock solution until the level of solution reaches the fill-line. Now your cell is ready to be wired.

Obtain a DC ammeter and two leads with alligator clips. Connect your electrolysis cell and the ammeter in series and ask your lab instructor to check your wiring before going further. When your wiring is approved by your lab instructor, make the final connection to the DC power supply. DC power is available at your lab table at outlets that require a special three-pronged plug or from a battery pack.

B. The Effect of Electrode Distance (individual work)

Position the electrodes so that they are directly opposite one another. Measure and record the distance between the electrodes in this position. For best results be sure that the electrodes are stabilized—not moving—before you turn on the current. Turn on the ammeter and record the current. Repeat the procedure with solution in the second beaker size. Compare your measurements with those of your partner. Record the electrode distance and the current generated in all four of the different-sized beakers.

C. The Effect of Electrode Position (individual and team work)

Individual work: With the electrodes in place in the smaller beaker and the ammeter turned on, slowly slide one electrode around the beaker until the two electrodes are almost touching. Record the highest current attained with the electrodes in closest proximity. Compare your findings with those of your partner. Record the initial and final currents measured by you and your partner along with the approximate volume of solution in the beakers.

Team work: Each partner should do the work involved with his or her setup. The other partner will control the ammeter and record current readings. Turn off the ammeter and remove one of the electrodes. Hold the electrode so it is directly opposite and parallel to the other electrode and submerge it in the solution. Turn the ammeter on. Hold the electrode as still as possible and keep the electrode positions parallel. Record the current. Slowly move the electrode forward until the two electrodes are *almost* touching. Record the highest current attained. Repeat the procedure with your partner's setup.

---

* These experiments are modeled on a procedure reported by Joseph Wang and Carlo Macca; *J. Chem. Ed.*, **73**, *8*, **1996**, 797–800.

D. The Effect of Concentration Changes (individual work)

Reposition the electrodes on the beaker in a fixed position directly opposite one another. Pour out approximately half of the solution from the smaller beaker and replace it with distilled water, filling to the mark on the electrode. Swirl gently to mix; then measure and record the maximum current generated for the diluted solution. Repeat the procedure for a second dilution of the same solution (i.e., pour out half of the solution, fill to the mark with distilled water, and record the maximum current).

Part III: The glucose sensor

A. Preparation of the Monitor (individual work)

Closely examine the glucose monitor and test strips. Follow the directions given with the kit for coding the monitor for the test strips you will use. Turn the monitor on. Be sure that the "low battery" signal is not showing. If necessary, replace the batteries or obtain a new test kit. Obtain a small amount of both the "high" and "low" glucose stock solution. Check that your monitor is working correctly by inserting a test strip and placing a drop of the low-concentration glucose stock solution on the test spot. The monitor should show a low reading. Carefully blot the test strip dry and repeat the testing with a drop of the high concentration glucose stock solution. The monitor should show a high reading. If you do not get appropriate glucose readings for this test, turn the monitor off. Then repeat the test with a new test strip. If you still do not get the desired result, seek further instructions from your lab instructor.

B. Preparation of Standard Solutions of Sample (individual work)

Prepare three or four standard solutions of glucose by diluting the sample with buffer stock solution. Suggested dilution ranges for various samples are

Sample Type	Dilution Range
Sunkist	1:75–1:100
Coke	1:50–1:75
Gatorade	1:20–1:40

For a 1:75 dilution, add 74 mL of buffer to 1 mL of Sunkist. For a 1:85 dilution, add 84 mL of buffer + 1 mL of Sunkist and so on. You will use the diluted samples to generate a calibration curve for your sample type, for example, Sunkist. Then you will use the calibration curve to determine the concentration of an unknown (Sunkist) sample. Save these standard solutions for use in the last part of this experiment.

C. Preparation of a Calibration Curve (individual work)

Obtain an unknown for your sample type. Take a new test strip out of the box and insert it into the monitor. You will use a new test strip to obtain a glucose reading for each of the diluted standard solutions prepared earlier and for an unknown solution containing your sample type. Using one of the diluted standard solutions you prepared, place a drop on the test spot and record the monitor reading. The test spot must be completely covered to obtain an accurate reading.

After the reading, turn off the instrument and carefully dry the test spot by gently pressing a clean filter paper or paper towel over it. Do not rub. Mark this strip and set it aside. Turn on the instrument, insert a new strip, and place a drop of the second diluted stock solution on the test spot. Record the monitor reading. Repeat this process for the third diluted stock solution.

Check that the concentration of the unknown sample falls within the calibrated range of values; use old strips for this. Carefully dilute the unknown if necessary, record the dilution made, and prepare to measure its concentration. With a new strip, measure and record the monitor reading for the unknown.

QUESTIONS TO ANSWER IN YOUR REPORT

Team Discussion

Part II: Investigation of an Electrolysis Reaction

1. What happens to the voltage when the electrodes are moved relative to one another?
2. What happens to the voltage when the electrodes are close together? When the electrodes are far apart? Is there any relationship or is the difference random?

Part III: The Glucose Sensor

1. What is the concentration range for the glucose tester?
2. How closely do your results agree with your partner's results? Explain.

Individual Report

1. Prepare a table of the concentrations of your diluted standard solutions and the monitor readings obtained for each standard solution. Plot a calibration curve of standard concentrations against monitor readings or calculate the least squares calibration equation. Determine the sample concentration from the calibration graph or equation. Remember to multiply by the dilution factor if you diluted the unknown solution.
2. Include the table of data with your calibration graph.
3. Submit a sample calculation that shows how you determined the concentration of your unknown solution.

EXPERIMENT GROUP H

HEAT CAPACITY: DESIGN A FIREPROOF SAFE

Sophisticated materials-testing labs work on the same calorimeter principles covered in general chemistry.
Courtesy of John Gadja, Construction Technology Laboratories.

PURPOSE   This module is designed to explore the role of heat capacity as a factor in choosing a material with special heat requirements. Heat capacity is an extensive property of a substance, one that depends on the amount of substance present. Different substances respond differently to heat. Heat capacity measures the ability of a substance to hold or store heat. Some substances require only a little heat to achieve a temperature change; other substances require considerable heat to raise their temperature even one degree. Heat capacity is a measure of this property and therefore is a useful tool in helping engineers choose appropriate materials for products that have special heat requirements. When they are heated under the same conditions, substances with small heat capacities will reach a high temperature sooner than will substances with large heat capacities.

SCHEDULE OF THE LABS   *EXPERIMENT 1: Skill Building Lab: "The Heat Capacity of Some Solid Elements"*

1. Construct a calorimeter and determine its heat capacity (individual work).
2. Determine the specific heat capacity of copper.
3. Identify an unknown metal on the basis of its specific heat capacity.
4. Use uncertainty analysis to determine an approximate range of acceptable values for the specific heat of both metals.

EXPERIMENT 2: Foundation Lab: "Thermochemistry of Reactions in Solution"

1. Measure the heat evolved in a chemical reaction (individual work).
2. Compare ΔH with ΔE for this reaction (group work).

3. Use uncertainty analysis to determine an approximate range of acceptable values for the heat of reaction.

EXPERIMENT 3: Application Lab: "The Fireproof Safe"

1. Determine the specific heat for zinc, cork, and concrete (group work).
2. Choose the best interstitial material for the fireproof safe from a comparison of graphed data for copper, zinc, cork, and concrete.
3. Justify your choice of material by comparing the calculated temperature of the safe (after half an hour in the fire!) for all the materials measured (individual work).
4. Select the filler material that best meets the requirements for the fireproof safe (group work).

SCENARIO Thermal insulation is a big business! Thermal insulation provides a barrier that reduces the amount of heat that can be transferred between the insulated substance and its surroundings. In the summer months, we are likely to consider thermal insulation in terms of the coolness of our houses, our lunches, or our beverages; in the winter months, we appreciate the heat that is retained by thermal insulation.

The choice of material used for thermal insulation depends on many factors. Resistance to heat flow is a primary consideration but not the only one. Other considerations are ease of handling, resistance to compression and corrosion, toxicity, in-fire performance, and cost. Building contractors purchase thermal insulation in several forms: rigid sheets that can be nailed into place, loose fill insulation that can be poured or blown into a space, and blanket insulation that can be unrolled and put into place.

When the last flames of the great Chicago fire of 1871 finally flickered out and people were able to return to their homes and offices, they were amazed at what they found. The heat of the fire had melted even metal and twisted it into grotesque globular shapes. Today, the steel girders used in building construction are encased in ceramic bricks or in concrete specially formulated to resist heat.

Inside the buildings, furnaces and chimneys are lined with fire-clay bricks that are made to withstand very high temperatures. Such materials are called refractory materials or simply refractories. Temperatures within a burning building may be in excess of 820°C. Fire-clay bricks can withstand temperatures to between 1500–1750°C, and some superrefractory materials such as silicon carbide are useful for temperatures up to 2204°C.

You have just been hired by a major safe manufacturing company that wants to branch out into the design and manufacture of fire-resistant boxes. Many businesses and homeowners are interested in such fireproof safes and use them to store valuable papers and jewelry that they want to protect from destruction by fire. Your first assignment in your new job is to develop a workable blueprint for the manufacture of small, fire-resistant boxes. What do you need to consider?

To answer this, you first must consider the purpose of making a safe fireproof. Obviously, you don't want the materials inside to melt or to burn. This means you must control the flow of heat between the fire on the outside and the contents of the safe. Most fireproof safes are constructed by putting a smaller box inside a larger box. The filler (or interstitial) material between the two boxes must prevent the heat of the outside fire from damaging the contents inside the safe. The ultimate question, then, is to identify a reasonable material for this purpose.

You will have to examine the major factors that govern the amount of heat that can pass through the filler material as well as other manufacturing considerations, such as the cost of such materials. You must also develop a method or technique that can be used to measure the flow of heat into or out of a system. Then you must test all of the materials available to your lab group in order to determine an appropriate filler material. You will base your decision on the information you obtained from calorimetry experiments, from your analysis of the factors important to this problem, and by comparing the temperatures inside the safe (using several different filler materials) after 30 min in a fire.

As preparation for this task, you will construct and use a calorimeter to measure the specific heats of some metals. This will familiarize you with the calorimeter setup and give you practice making the measurements necessary for heat calculations. Experiment 2 will provide a different experience of heat measurement. You will measure the heat evolved in a chemical reaction that takes place in aqueous solution. Some industrial applications require the containment of generated heat inside a closed system. You will apply all that you have learned about heat measurement in the third lab, when you test various filler materials for the fireproof safe. In this application, the filler materials must keep generated heat outside the system.

Experiment 1
SKILL BUILDING LAB: THE HEAT CAPACITY OF SOME SOLID ELEMENTS

Pre-Laboratory Assignment   **Due Before Lab Begins**

NAME: _____

Complete these exercises after reading the experiment but before coming to the laboratory to do it.

1. One of the factors engineers must consider is the validity of any assumptions made while they perform their jobs. In this experiment, several temperature assumptions are made. List two of these assumptions and explain why you think they are reasonable assumptions to make.

2. a. Write the mathematical expression of the Law of Dulong and Petit.

   b. Using the Dulong and Petit approximation, estimate the specific heat capacity, c_s, of manganese.

   c. Calculate the percent error resulting from this approximated value of c_s. Do this by comparing it with the known value for manganese that is listed in *The CRC Handbook of Chemistry and Physics*, 0.114 cal g^{-1} K^{-1}. (Remember, 1 cal = 4.184 J.)

   d. Derive the Law of Dulong and Petit using the following relationships: $q = mc_s\Delta T$, $q = nc_p\Delta T$, and M (the molar mass) $= m/n$.

3. What is the primary function of a calorimeter for this experiment?

4. During an experiment you perform many different procedures and the final results of your experiment depend largely on how well you carry out these procedures as well as on the design of your lab apparatus. Decide whether the following conditions will have a large or a small effect on your experimental results. Give a reason for your answer.

   a. The calorimeter lid has a large hole in it.

   b. Hot water is transferred along with the metal into the calorimeter water.

   c. The initial temperature of the cold water is 28°C.

   d. You forget to put the lid on the calorimeter.

   e. You use an alcohol thermometer instead of a mercury thermometer.

   f. You record the final temperature before thermal equilibrium is established.

5. Suppose a hot (89.5°C) piece of copper metal ($c_s = 0.385$ J g^{-1} K^{-1}) with a mass of 2.55 grams is put into 50.0 g of water ($c_s = 4.184$ J g^{-1}K^{-1}). Calculate the final temperature of the water if its initial temperature is 25.7°C.

6. What is the general method used in your lab to heat water? List two safety precautions you must heed during the heating process.

Experiment 1
SKILL BUILDING LAB: THE HEAT CAPACITY OF SOLID ELEMENTS

BACKGROUND

Heat capacity measures the amount of heat needed to raise the temperature of a substance one degree. In other words, heat capacity measures the ratio of heat to temperature change, $C = q/\Delta T$, where C is the heat capacity, q is the heat, and $\Delta T = T_{\text{final}} - T_{\text{initial}} = T_f - T_i$. Substances with large heat capacity values have large values of the ratio $q/\Delta T$. They can be thought of as good "heat storing" substances. Substances with smaller heat capacity values are less good at storing heat. Heat capacity is an important property to industries whose livelihood depends on the ability to control heat transfer.

The heat capacity of an object depends on the size (mass) of the object. It is an extensive property. For example, 2000 kg of water has a greater heat capacity than 20 kg of water. In other words, 2000 kg of water can store more heat than 20 kg of water. To directly compare the heat-storing capability of two or more substances we use **specific heat capacity** (c_s), which is the heat capacity *per gram* of a substance, $c_s = q/(m\,\Delta T)$. The specific heat capacity is used in the calculation of heat,

$$q = mc_s, (T_f - T_i) = mc_s\,\Delta T$$

where m is the mass of the substance in grams and c_s has units of J g^{-1} K^{-1}

For mole amounts, we use the **molar heat capacity** (c_p). Molar heat capacity is the amount of heat needed to raise the temperature of one *mole* of substance one degree. Then the molar heat capacity is $c_p = q/(n\,\Delta T)$ and heat is calculated as $q = nc_p\Delta T$, where n = number of moles of a substance and c_p is the molar heat capacity with units of J mol^{-1} K^{-1}.

When expressing the absolute temperature of a system, we usually *must* convert to the Kelvin scale. However, when measuring temperature **change,** we may use either the Kelvin or the Celsius scales. The Kelvin and Celsius temperature scales are different but, because the size of the degree unit in these two scales is the same, ΔT has the same numeric value in both temperature scales. A temperature change from 32.4 to 26.2°C gives $\Delta T = 6.2$°C. The corresponding change on the Kelvin scale is from 305.6 to 299.4 K; $\Delta T = 6.2$ K.

CALORIMETRY

Heat capacity values can be experimentally determined using calorimetry methods. Calorimetry experiments use initial and final temperature measurements to determine the heat generated or absorbed by some change. The change may be as simple as the heat transfer between a hot piece of metal and cold water, or it may involve measuring the heat of a chemical reaction. To determine ΔT in a calorimetry experiment, it is necessary to control the heat transfer by isolating the system of interest in a well insulated container called a calorimeter.

In thermodynamics we define the universe as consisting of two parts: 1) a particular system of interest (usually just called the system) and 2) everything else outside the system (usually called the surroundings). Energy can be transferred between the system and its surroundings but the total amount of energy distributed between the two remains constant. In other words, $\Delta E_{\text{univ}} = \Delta E_{\text{sys}} + \Delta E_{\text{surr}} = 0$. For this experiment, the surroundings will be the calorimeter; the system, then, is the contents of the calorimeter. We are only concerned with thermal energy monitored by the variable heat, q, so $q_{univ} = 0 = q_{sys} + q_{surr}$.

Calorimeters can be constructed using a variety of designs. The Styrofoam® cup containing your hot morning coffee is a type of "calorimeter" designed to minimize the transfer of heat from the coffee to the surrounding air. Your calorimeter will have a lid so that essentially all of the heat transferred stays inside the calorimeter. Since the calorimeter is the only component of the surroundings, we assume that $q_{surr} = q_{cal}$.

The amount of heat actually absorbed by the calorimeter can be determined experimentally by measuring the heat transfer that occurs when hot water is added to cold water inside the calorimeter. Theoretically, the heat lost by the hot water should equal the heat gained by the cold water. This is the Law of Conservation of Energy. Any discrepancy in the two values is due to the heat absorbed by the calorimeter. The amount of heat absorbed by the calorimeter is related to the heat capacity of the calorimeter and is expected to be the same for similar systems. The heat capacity of the calorimeter is symbolized as C_{cal}. C_{cal} has units of J K^{-1}.

The overall system is adiabatic ($q_{univ} = 0$), so the calorimeter constant is calculated by

$$q_{univ} = 0 = q_{sys} + q_{surr}$$

$$q_{surr} = q_{cal}$$

$$q_{sys} = q_{cw} + q_{hw}$$

$$q_{cw} = c_{cw}m_{cw}\Delta t_{cw}$$

$$q_{hw} = c_{hw}m_{hw}\Delta t_{hw},$$

$$q_{cal} = -(q_{cw} + q_{hw}) = C\Delta t_{cal}$$

q_{cw} = heat change for the cold water
q_{hw} = heat change for the hot water
q_{cal} = heat change for the calorimeter.

The specific heat of water is: $c_{cw} \approx c_{hw} \approx 4.184$ J g^{-1} K^{-1}.

To determine q_{cal}, you need to evaluate q_{cw} and q_{hw}. For the same calorimeter, C_{cal} may be considered constant when used with comparable masses of water. However, ΔT will most likely vary from experiment to experiment so q_{cal} must be recalculated for each experiment.

In this experiment, you will also measure the temperature of cold water before and after some hot metal is added to it. When two substances having different temperatures are placed into close physical contact with one another, heat will flow from the hotter substance into the colder substance until both substances arrive at the same temperature. They will attain "thermal equilibrium." A simple example of this is the immersion of a hot metal into a beaker of cold water. The transfer of heat from the metal to the water can be followed by monitoring the temperature change of the water. The water temperature measured when the metal–water system has reached thermal equilibrium is the final temperature for both substances. Measurement of this temperature change allows one to calculate the heat transferred from the metal to the cold water and the calorimeter. Heat is calculated as the product of mass, specific heat capacity, and temperature change: $q = mc_s \Delta T$

At thermal equilibrium, the heat given off by the metal and the heat absorbed by the water are equal but are given opposite signs by convention. In thermodynamics, whatever leaves the system is considered a negative quantity while that which is absorbed by or added to the system is considered to be positive in sign. In this experiment, heat leaves the metal and is absorbed by the cold water and the calorimeter.

Since your overall system is adiabatic, we can write

$$q_{univ} = 0 = q_{sys} + q_{surr}$$

$$q_{sys} = q_{cw} + q_{metal}$$

$$q_{surr} = q_{cal}$$

$$q_{cw} + q_{metal} + q_{cal} = 0$$

$$q_{cw} + q_{cal} = -q_{metal}$$

$$(mc_S \Delta T)_{cw} + C\Delta t_{cal} = -(mc_S \Delta T)_{metal}$$

Note that in this case the values for ΔT will be different for the metal and the cold water.

Substances attain thermal equilibrium when their temperatures become equal after some period of close physical contact. A solid substance can attain thermal equilibrium with a second substance (such as water) in many different ways. In the first part of this experiment, the solid substance is a metal and is not soluble in water. To bring the metal to the temperature of the hot water, you need an experimental design that will keep the metal dry and allow you to transfer the heated metal quickly (and safely) to the beaker containing cold water. Large pieces of metal could reasonably be contained in a heat-resistant zip-lock plastic bag. Small pellets of metal could be heated inside a test tube immersed in hot water. With the passage of 7 to 10 min, it is reasonable to assume that the heat from the water bath has been transferred to all of the metal.

This experimental setup has the additional advantage of ensuring that the final temperature measured is due solely to the transfer of heat from the metal to the cool water. By initially keeping the metal sample dry, no water from the hot water bath is transferred into the calorimeter beaker along with the heated metal.

The experimental setup for this experiment is straightforward. It involves two beakers of water, one that is used to heat a known mass of the metal to some measured, elevated temperature, and a second, cooler sample of water into which the hot metal will be put. The amount of water in the first beaker is unimportant because it is used merely to heat the metal. It is crucial, however, to measure the mass of water in the second beaker because this mass enters into the calculation of heat. The initial temperature of the hot metal will be assumed to be the same as that of the hot, boiling water. The initial temperature of the cold water will be measured before the addition of the hot metal. The final temperature of both the metal and the cold water will be the same—the highest water temperature attained after the addition of the hot metal to the cold water. With all of these variables measured, you can calculate the specific heat of the metal.

Throughout the experiment and the calculations, remember three important points:

1. The major assumption is that there is no heat flow outside of the calorimeter.
2. Neither the heat capacity nor the specific heat of any component changes either with temperature or with time.
3. Heat absorbed is a positive quantity, heat given off is a negative quantity. Or, if something "gives off" heat, then its "absorbs" a negative amount of heat.

Dulong and Petit Approximation

In 1819, Dulong and Petit observed an interesting trend in the measured *molar* heat capacities of some solid elements. They found their experimental values all fell within the range of values from 22 to 28 J mol^{-1} K^{-1}. Using an average value of 25 J mol^{-1}

K^{-1}, Dulong and Petit derived an approximate relationship between *specific* heat capacity and molar mass,

$$c_s M \approx 25 \text{ J mol}^{-1} \text{ K}^{-1}$$

where c_s is the specific heat and M is the molar mass of the substance. This relationship allows one to approximate the molar mass of an unknown solid element by measuring its specific heat! It is called the Law of Dulong and Petit. In this experiment you will measure the specific heat capacity of an unknown element and identify it from the calculated molar mass using the Law of Dulong and Petit.

Caution: This experiment necessitates the use of hot water. The two most common methods of heating water in the lab use either a bunsen burner or a hot plate. If your lab uses open flames, such as those produced by a bunsen burner, be careful that no part of your body (including your hair) or your clothing comes into contact with the fire. You must wear goggles at all times in the lab. You must also be careful that other people in your lab are not working with flammable chemicals. Flammable chemicals must not be opened or used in the vicinity of an open flame.

If your lab uses hot plates for heating purposes, follow the directions of your lab instructor about the location and usage of the hot plates. Most hot plates are equipped with a light that is on when the hot plate is on. Hot plates may appear the same to you whether they are turned on (and hot) or turned off (and cold). Observe proper precautions when working in the vicinity of the hot plate.

PROCEDURE

For group discussion and data analysis purposes, this experiment is best done in groups of two. Your group should collaborate on the construction of the calorimeter, including all mass and temperature measurements. All measurements must be carefully made and recorded if you are to have successful results. The aim is to determine a specific heat capacity for copper that closely approaches the value reported in the literature. Begin now by preparing a table that specifies the measurements you will contribute to the group.

Your notebook should include the masses and temperatures needed for the calculation of specific heat capacity.

Engineers do many experiments during the course of a day, a week, or a year. They must be prepared to discuss, analyze, or report on any one of these experiments at any time. Part of their job is to carefully record the laboratory setup, conditions, assumptions, and all data that might help them fulfill these responsibilities. Measurements, changes in procedure, accidental spillage, and all other details that may influence experimental results should be recorded in your notebook so that everyone in the group can know what happened. Take time now to prepare your notebook; indicate the method used, sketch a picture, and keep track of problems and difficulties so that you can avoid them in the future.

Part I: Construction of a Calorimeter (Individual Work)

There are several kinds of calorimeter you may use in this experiment. One is a simple Styrofoam cup with a foil cover. It has the advantage that it is easy to construct and use. However, the light weight of the Styrofoam means that you must be especially careful to avoid tipping it over.

To use Styrofoam cups, obtain two cups to nest one inside the other and a 3- × 3-in. square piece of foil to form a cover for the nested cups. Carefully mold the foil to the top of the cup assembly so that it can be removed and replaced easily. Pierce a small hole in the center of the foil just large enough to accommodate a thermometer. The nested Styrofoam cups can be stabilized by resting them inside a glass beaker.

A sturdier calorimeter may be made by placing a 250-mL beaker inside of a larger beaker or inside of a special metal can. The 250.0-mL beaker is used as the inner cup of the calorimeter. If the outer part of the calorimeter is a larger beaker, you can use paper toweling as insulation between the two beakers. There are also metal cans available to use as the outer part of the calorimeter. In this case, the 250.0-mL beaker is supported by a metal ring that fits inside the can. When properly assembled, the beaker is suspended by the insulating metal ring so that no part of the beaker touches the can. Further insulation can be provided by wrapping the inner beaker with paper towels. A lid can be fashioned from a 5- × 5-in. square of cardboard. Cut a small hole into the center of the cardboard for the thermometer. Wrap a rubber band around the thermometer to position it in place so that it does not touch the bottom of the calorimeter beaker.

Yet another "calorimeter" is a small thermos. In fact, any insulated container makes an acceptable calorimeter if it minimizes the transfer of heat between the inside reaction container and the atmosphere. Follow the direction of your laboratory instructor in your choice of calorimeter materials.

Part II: Measurement of the Heat Capacity of the Calorimeter (Individual Work)

Weigh out two samples of deionized water. An appropriate amount of water for this experiment is about 50 g. The mass can be any value within 10% of this. Record all masses to a precision of ±0.01 g. Put one water sample into the calorimeter beaker and measure its initial temperature. Measure the thermometer readings at eye level and carefully estimate the temperature to within ±0.2°C. Heat the second sample of water until its temperature is about 10 to 15° higher than that of the cold water. Record the initial temperature of the warm water. When you are ready, quickly pour the warm water into the calorimeter beaker with the cooler water, replace the lid, and monitor the temperature as it changes. Record the highest temperature attained by the water mixture in the calorimeter. The system is assumed to be at thermal equilibrium when the change of temperature with time becomes very small. There is only one final temperature; it is the same for both the hot and the cold water.

Repeat for a total of two trials.

During the laboratory period it is important that you detect any mistakes in procedure or technique before going on to the next part of the experiment. Calculate the heat capacity of your calorimeter for each trial. The two values should be within 3% of each other. If not, repeat the determination until you can get the values to be consistent. Use an averaged value of C_{cal} for the remainder of the calculations.

Part III: Obtain the Specific Heat Capacity of Copper (Group Work)

Preparation of the Calorimeter Water (pair work)

Place about 100 g of deionized water into the calorimeter beaker. (*Record* the precise mass in your laboratory notebook.)

Record the temperature of the water to ±0.2°C. This temperature is the *initial temperature* of the water in your calculations.

Heating the Sample (pair work)

Half-fill a test tube with an accurately measured mass of copper metal pellets. Heat the test tube and metal contents to thermal equilibrium in a hot water bath, being careful to keep the metal dry. However, to guarantee that all of the metal has reached thermal equilibrium, be careful to keep the level of the water surrounding the test tube above the contents of the test tube. Allow about 7 to 10 min for thermal equilibrium to be established. Then record the temperature of the water bath to the same degree of precision as you did for the cooler water in the calorimeter. This will be the initial temperature of the metal in your calculations.

When your group is ready, pour the hot metal into the water in the calorimeter. Do this transfer quickly but efficiently. Work together to take off and to replace the calorimeter lid. Carefully mix the contents of the calorimeter while you watch the thermometer. Note and record the highest temperature attained (to ±0.2°C). This number is the *final temperature* of the water in the calorimeter, the calorimeter, *and* the metal, as all three are in thermal contact.

Perform *three* experimental trials.

Part IV: Identification of an Unknown Metal (Group Work)

Repeat the procedure, of Part III with an unknown metal. Do *three* trials.

REPORT

1. Report the three values measured and the average value computed for the specific heat of copper. Calculate the percentage error between your averaged result and the reported value found in the *CRC Handbook of Chemistry and Physics* or your textbook.

2. Report the three values measured and the average value computed for the specific heat of your unknown metal. Use the Law of Dulong and Petit to find an approximate molar mass for your unknown. Finally, attempt to identify the metal by consulting the periodic table. Justify your choice.

3. *Include in your lab report* a calculation of the uncertainty propagated through the calculation of specific heat capacity for copper (see Introduction).

Experiment 2
FOUNDATION LAB: THERMOCHEMISTRY OF REACTIONS IN SOLUTION

Pre-Laboratory Assignment   **Due Before Lab Begins**

NAME: _____

Complete these exercises after reading the experiment but before coming to the laboratory to do it.

1. Define: heat capacity, molar heat capacity, and specific heat capacity.

2. A 25.00-g sample of water at a temperature of 99.6°C is added to 25.00 g of water at an initial temperature of 23.6°C. The temperature is recorded every 10 s. From the following data, calculate (a) the heat lost by the hot water, (b) the heat gained by the cold water, and (c) the heat gained by the calorimeter. (The specific heat capacity of water is 4.184 J g^{-1} K^{-1}.)

Time, s	Temp, °C	Time, s	Temp, °C	Time, s	Temp, °C	Time, s	Temp, °C
10	23.8	50	38.5	90	55.6	130	59.1
20	25.9	60	49.7	100	57.8	140	59.1
30	29.1	70	50.8	110	58.9	150	59.0
40	34.3	80	51.2	120	59.2	160	58.9

3. A substance is cooled from 74.2°C to 44.8°C. Calculate ΔT in °C. Calculate ΔT in Kelvin.

4. Name two assumptions concerning heat transfer that are validated because the system is adiabatic.

5. Calculate the change in internal energy (ΔE) that occurs in the combustion of 5.00 g of methanol, CH_3OH, at a temperature of 298 K. Methanol burns according to

$$CH_3OH\ (g) + \frac{3}{2}\ O_2\ (g) \rightarrow CO_2\ (g) + 2\ H_2O\ (g) \qquad \Delta H = -676.5\ kJ$$

6. Since it is important for you to measure temperatures accurately in this experiment, you may be using mercury thermometers. Why must you not use the thermometer as a stirrer in your calorimeter?

Experiment 2
FOUNDATION LAB: THERMOCHEMISTRY OF REACTIONS IN SOLUTION

BACKGROUND

The first law of thermodynamics tells us that the total amount of energy in the universe is constant. We usually think of this as the Law of Conservation of Energy. The interesting thing about energy is that it can take many different forms—heat, light, and electricity are all ways in which energy is manifested. Because energy can be transferred from substance to substance using many different vehicles, it is difficult to keep track of the energy involved in chemical processes.

In thermodynamics we define the universe as consisting of two parts: (1) a particular system of interest (usually just called the system) and (2) everything else outside the system (usually called the surroundings). Energy can be transferred between the system and its surroundings, but the total amount of energy distributed between the two is the same before and after any change occurs in the system. In a chemical reaction, the system of interest is the reaction itself and the surroundings are usually taken to mean the immediate surroundings, including the container and the surrounding air; in short, anything that touches the system. Ordinary chemical reactions allow us to measure some interesting thermodynamic quantities. For example, the internal energy of the system changes with each bond that is broken or formed. The change in internal energy (ΔE) is the sum of two terms, heat and work:

$$\Delta E = q + w$$

For experiments performed under ordinary laboratory conditions, the pressure is assumed to be constant. This is shown in measurements of heat by a subscript p, signifying heat at constant pressure (q_p). Heat at constant pressure is the same as the change in enthalpy, ΔH

$$q_p = \Delta H$$

For systems under normal laboratory conditions of pressure, then, we can measure ΔH quite simply by measuring q_p.

The change in internal energy (ΔE) is

$$\Delta E = q_p + w = \Delta H + w = \Delta H + P\,\Delta V$$

Here the work term simplifies to the volume expansion of a gas against the constant pressure of the atmosphere.

For chemical reactions that involve no expansion of gases,

$$\Delta E = \Delta H = q_p$$

For chemical reactions that generate a number of moles of gas that differs from that originally present in the reactants, however, the work term must be evaluated. If we assume that any gas generated by the chemical reaction behaves as an ideal gas, then the PV term can be evaluated by making appropriate substitutions using the ideal gas equation,

$$V = \frac{nRT}{P}$$

$$P \, \Delta V = P \, \Delta \left(\frac{nRT}{P} \right) = RT \, \Delta n$$

where $R = 8.314$ J mol^{-1} K^{-1}, and Δn = number of moles of gas (product) − number of moles of gas (reactants). This substitution allows a calculation of ΔE given ΔH, the temperature, and the chemical equation for the reaction.

Heat

Heat is a means by which energy can be transferred between a system and its surroundings. Some reactions require the addition of heat. These are called **endothermic** reactions. As endothermic reactions absorb heat, $q_{rxn} = \Delta H_{rxn} > 0$. Reactions that liberate heat to their surroundings are called **exothermic** reactions. In an exothermic reaction, $q_{rxn} = \Delta H_{rxn} < 0$. Although heat and temperature are not the same thing, there is a direct correlation between them. As heat flows into an object, the temperature generally increases; as heat flows out of an object, the temperature decreases. Hence, heat transfer can be monitored and measured by watching the temperature.

Thermochemistry is a branch of chemistry that studies heat transfer. For heat transfer to occur there must be thermal contact between the substances in the system. To measure the heat transferred during a chemical reaction, it is imperative that we prevent the heat from escaping or diffusing into the surrounding atmosphere before a proper measurement can be obtained. This is accomplished by using a calorimeter.

In this experiment, you will put a weighed amount of H_2O_2 solution into a calorimeter and measure the heat that is given off by the decomposition reaction of H_2O_2 occurring in the solution. You can assume that all of the heat transferred in the reaction stays inside the calorimeter (i.e., assume adiabatic conditions). The H_2O_2 solution contains a very small amount of H_2O_2 (1–5% by mass) and a predominant amount of H_2O (99–95% by mass). The heat liberated by the decomposition of H_2O_2 is absorbed by the solution water and by the calorimeter.

The Decomposition Reaction

We often use a dilute solution of hydrogen peroxide, H_2O_2, to thoroughly cleanse cuts and abrasions. It is a familiar first aid remedy. When H_2O_2 contacts your skin, it bubbles vigorously, cleansing away surface grit. Warmth is a secondary sensation in the cleansing process, because the reaction is exothermic. The bubbling action is due to the rapid decomposition of the hydrogen peroxide,

$$2 \, H_2O_2 \, (aq) \rightarrow 2 \, H_2O \, (l) + O_2 \, (g) \tag{1}$$

This decomposition reaction is the one that you will be examining today. The isolated decomposition of H_2O_2 occurs very slowly at room temperature. However, the reaction can occur very quickly in the presence of a suitable catalyst (a catalyst is a substance that can alter the rate of a chemical reaction without itself being permanently changed).

To better understand how a catalyst may work, let's look at another system. The decomposition of hydrogen peroxide can also be catalyzed by iron(II) ion. In acid solution, iron(II) ion is oxidized in the first step of a two-step reaction mechanism.

$$2 \, Fe^{2+} \, (aq) + 2 \, H^+ \, (aq) + H_2O_2 \, (aq) \rightarrow 2 \, H_2O \, (l) + 2 \, Fe^{3+} \, (aq) \tag{2}$$

This begins a cycle in which iron(II) ion is initially oxidized to iron(III); then the iron(III) ion is reduced back to iron(II) ion. Thus the initial iron(II) ion is regenerated, and the reaction proceeds quickly in this cyclic manner.

$$2\ Fe^{3+}\ (aq) + H_2O_2\ (aq) \rightarrow 2\ H^+\ (aq) + O_2\ (g) + 2\ Fe^{2+}\ (aq) \tag{3}$$

The overall reaction is the sum of these two steps and results in equation (1), the simple decomposition of hydrogen peroxide.

$$2\ Fe^{2+}\ (aq) + 2\ H^+\ (aq) + H_2O_2\ (aq) \rightarrow 2\ H_2O\ (l) + 2\ Fe^{3+}\ (aq) \tag{2}$$
$$\underline{2\ Fe^{3+}\ (aq) + H_2O_2\ (aq) \rightarrow 2\ H^+\ (aq) + O_2\ (g) + 2\ Fe^{2+}\ (aq)} \tag{3}$$
$$2\ H_2O_2\ (aq) \rightarrow 2\ H_2O\ (l) + O_2\ (g) \tag{1}$$

To catalyze the decomposition of H_2O_2 for your experiment, you will use an enzyme called catalase that is derived from bovine (cattle) liver. Catalase is present in many different organisms, and it functions to protect the organism from the destructive—even toxic—reaction between peroxide and animal tissue. The human variant of catalase is the substance responsible for the bubbling that occurs when hydrogen peroxide is poured on an open wound.

In this experiment, you will use several different concentrations of hydrogen peroxide solution. The concentrations will be % (by mass) solutions. Because hydrogen peroxide is the active ingredient in this reaction, the heat liberated by the reaction is directly proportional to the amount of H_2O_2 present in the sample.

Caution: Although the solutions of hydrogen peroxide you will be using are very dilute, you should always observe proper laboratory safety precautions when handling chemicals of any kind. Avoid chemical contact with your skin and especially with your eyes. Safety goggles should be worn at all times in the lab.

PROCEDURE

Part I: Formation of Groups

For group discussion and data analysis purposes, this experiment is best done in groups of three. Each individual in the group is responsible for building and calibrating his or her own calorimeter and for measuring one of the three concentrations of H_2O_2 available in the lab. There will be two samples of each concentration to measure. At the completion of the experiment, the group will pool the data for the different concentrations of hydrogen peroxide and compare the heats of reaction obtained.

Part II: Construction of a Calorimeter (Individual Work)

Each member of the group should construct and measure the heat capacity of a calorimeter. It is necessary to determine C_{cal} for each new calorimeter you construct and use. (If necessary, refer to the directions given in experiment 1 in this Group.)

Caution: If your calorimeter is made of a beaker inside of a metal can, be sure to put the H_2O_2 solutions into the beaker and *not* into the metal can.

Part III: Heat of Reaction of Hydrogen Peroxide (Individual Work)

Each member of the group should work with a different concentration of hydrogen peroxide and should do two trials—one with approximately 50 g of solution and one with 75 g of solution. The concentration of the peroxide solutions will be expressed as mass percentage. Obtain the mass of the H_2O_2 solution. Make all mass measurements to ± 0.01 g.

Assemble the calorimeter with the H_2O_2 solution, cover, and insert the thermometer. Wait until the temperature stabilizes. Record the temperature to ±0.2°C.

When you are ready, add three to four drops of the catalase solution. Immediately replace the cover and thermometer in the calorimeter, and then swirl gently to mix well.

 Caution: Do not use the thermometer to stir the solution as it is fragile and may break. Instead, rotate the calorimeter in small circular motions so that the solution inside is gently mixed.

Watch the thermometer until the temperature reaches a maximum. The reaction occurs rapidly (in 2–4 min) if the catalase is freshly prepared. Record the maximum temperature attained. Repeat the reaction using the same calorimeter. This time, use a different mass of the peroxide solution.

Analysis of Data

The analysis of the data for this section is similar to the previous. Once again, the system is adiabatic. But now we can write a heat change associated with the reaction, q_{rxn}:

$$q_{univ} = 0 = q_{rxn} + q_{cal} + q_{H_2O}$$

Here q_{rxn} is the heat liberated by the decomposition of hydrogen peroxide, q_{cal} is the heat absorbed by the calorimeter and q_{H_2O} is the heat absorbed by the water in the solution. Assume this is the mass of the entire peroxide solution. You have experimental data with which to calculate q_{cal} and q_{H_2O}. Assume the specific heat of water in the solution is the same as that of pure water, 4.184 J g^{-1} K^{-1}.

All of the heat generated or absorbed by the reaction must be offset by heat absorption or liberation by the calorimeter and its contents. Therefore, solving for q_{rxn} we get

$$q_{rxn} = -(q_{cal} + q_{H_2O})$$
$$q_{rxn} = -(C \, \Delta T)_{cal} - (c_s m \, \Delta T)_{H_2O}$$

You know the mass of the hydrogen peroxide solution. The initial and final temperatures of the calorimeter and of the hydrogen peroxide solution are the same; the two are in contact at the start of the run. Remember that the heat capacity of the calorimeter is C_{cal} not q_{cal}. Once you know q_{rxn} you can calculate the molar heat of reaction, ΔH_{rxn}:

$$\Delta H_{rxn} = q_{rxn}/n$$

where n is the number of moles of H_2O_2 actually used in each experiment.

REPORT

Calculations

1. Determine C_{cal}, the heat capacity, for your calorimeter. Use an averaged value of C_{cal} for the rest of the calculations.
2. Calculate q_{rxn} for the concentration of peroxide you used. Do a calculation for both sample sizes. Express your answers in kJ.

3. Calculate ΔH_{rxn} for each solution you measured. Express your answers in kJ/mol.

4. Use standard heat of formation values (ΔH_f°) obtained from tables to calculate the theoretical value for the heat of reaction for $2\ H_2O_2 \rightarrow 2\ H_2O + O_2$. Express your answer in kJ/mol.

5. Compare your experimental value of ΔH_{rxn} with the theoretical value just calculated to determine the % error in your measurements.

Group Work

1. Prepare two tables to summarize the data obtained by your group. Each group member should contribute the following information for this table: (a) Table 1 should contain the % by mass concentration of the samples you measured, the actual mass of H_2O_2 used in your samples, and q_{rxn} for each sample; (b) table 2 should contain the % by mass concentration of the samples you measured, ΔH_{rxn} for these samples, and the % error. (See calculations 2, 3, and 5.)

2. Discuss the contents of these tables as a group, and formulate a group interpretation based on the values of q_{rxn} obtained (table 1) and of ΔH_{rxn} obtained (table 2).

3. Discuss the possible source of errors in this experiment. Make a list of three of these errors and include a prediction of how each error might affect the final result. (High result? Low result? No effect?)

4. For reactions in which the pressure or volume do not change, the Law reduces to $\Delta E = \Delta H_p$, but for reactions that generate a gas, this is not true. Then ΔE is calculated using the full equation, $\Delta E = \Delta H - P\,\Delta V$. Calculate ΔE for your reaction when T has a value of 313 K. Remember to make the appropriate substitution for PV.

5. Each member of the group should use uncertainty analysis (described later) to calculate the uncertainty in q_{rxn} due to the uncertainty in q_{H_2O}. (This is only part of the calculation you would have to do to find the total uncertainty in q_{rxn}, but it will serve as an example of the calculation.)

Uncertainty Analysis

This is a more rigorous mathematical treatment of the propagation of uncertainty in calculations than that discussed in the first experiment. Your instructor will tell you whether or not to do this part of the analysis.

1. To find the uncertainty in ΔT (which is an addition/subtraction procedure), the uncertainty is found by $\sqrt{e_1^2 + e_2^2}$, where e is the absolute uncertainty in each measurement. Each temperature can be read to $\pm 0.2°C$. This is the absolute uncertainty in the measurement. For ΔT, then, the uncertainty is calculated as $\sqrt{0.2^2 + 0.2^2} = \sqrt{0.04 + 0.04} = \sqrt{0.08} = 0.2_8$. The subscripted 8 is carried through the calculation for final rounding-off purposes. This uncertainty in ΔT will be used in the multiplication of "$mc_s\,\Delta T$" in the next step.

2. For multiplication and division procedures, we consider the percent error which relates the error or uncertainty to the actual measurement itself. For example, if the mass recorded is 24.20 ± 0.01 g, the ratio of error in mass to the mass is $(0.01/24.20) \times 100 = 0.041\%$. For the sake of simplification, let's call this e_1. We will find the % error or all mass and temperature measurements in the equation, $q_{H_2O} = (mc_s\,\Delta T)_{H_2O}$. We will assume the % error in the heat capacity of water is negligible and ignore it for this calculation. We will call each of these values, e_2, e_3, and so on.

For example, suppose the mass of hydrogen peroxide solution is 96.40 ± 0.01 g and $\Delta T = 16.4 \pm 0.2_8°C$. Then the % uncertainty in $m_{H_2O} = 0.01/96.40 \times 100 = 0.010\%$;

the % uncertainty in $\Delta T_{H_2O} = (0.28/16.4) \times 100 \times 1.7\%$; and the propagation of this uncertainty in the calculation is $\sqrt{0.01^2 + 1.7^2} = 1.7\%$.

The absolute error or uncertainty in our calculated value of q_{H_2O} is found by multiplying the % error in q_{H_2O} by the calculated value of of q_{H_2O}. In this example, the % error is 1.7% and the calculated measurement has a value of $(96.40 \text{ g})(4.418 \text{ J g}^{-1} \text{ K}^{-1}))(16.4 \text{ K}) = 6984.7 \text{ J}$. Then 1.7% of 6984.7 is 118.7. The uncertainty in this answer is shown by writing 6985 ± 119. In this example, the uncertainty in the calculated value reflects the uncertainty contributed to q_{rxn} by uncertainty in q_{H_2O}.

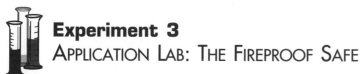

Experiment 3
APPLICATION LAB: THE FIREPROOF SAFE

Pre-Laboratory Assignment   **Due Before Lab Begins**

NAME: _____

Complete these exercises after reading the experiment but before coming to the laboratory to do it.

1. A fireproof safe is made of two thin metal boxes, one inside the other, with equal spacing between each pair of adjacent surfaces. Calculate the volume *between* the two boxes given the following dimensions. Include units.

    smaller box (25 m × 20 m × 15 m)
    larger box (32 m × 27 m × 22 m).

2. Calculate the mass of lead in a volume of 126 cm^3. The density of lead is 11.35 g cm^{-3}.

3. The exponential term in the calculation of inside temperature for a safe surrounded by fire contains several variables. (A discussion of the variables appears in the section titled "Group Report" at the end of this laboratory procedure.) The units for these variables must be such that they will divide out, leaving no units in the exponential term. Show this is true for the exponential term, $-\dfrac{hA_s}{c_s m}\, t$.

4. Suppose you spill the boiling water (used in this experiment to heat irregular solid substances) on your arm. What first aid procedure would you use to control the damage to the burned arm?

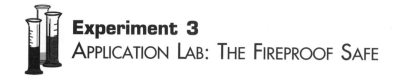

Experiment 3
APPLICATION LAB: THE FIREPROOF SAFE

BACKGROUND

In this experiment you will examine the heat capacities of three other substances that might be good filler materials for the safe: zinc, cork, and concrete. The shapes of these materials force us to explore other methods of heating them to attain the initial high temperature required. None of these materials will fit inside a test tube. Although we might consider direct submersion into the hot water for the piece of zinc, direct submersion introduces some degree of error if any amount of the hot water is transferred to the cold water container along with the metal. Direct submersion is a poor choice for either the cork (which is porous and absorbs water) or the concrete (which is a mixture and tends to fall apart in the hot water bath). A heat-resistant plastic wrap or bag would be a good alternative to the test tube we used for metal pellets. The plastic material allows close thermal contact between the solid and the hot water while preventing the solid from becoming wet.

By the end of this experiment, you will be asked to make a choice of filler materials based on your experimental results for the specific heat capacity of the candidate materials. The choice may depend on a variety of properties that we didn't test. Are there other factors you should consider in your choice of material? Suppose the only important property to consider is thermal conductivity. Thermal conductivity measures the transfer of heat by conduction across a solid. How do our materials compare in their ability to conduct heat? Instead of doing the experiment, we will look up these values in *The CRC Handbook of Chemistry and Physics*. The thermal conductivity of cork is 0.04 cal s^{-1} cm^{-2}. The thermal conductivity values of concrete, copper, and zinc are 1.4 cal s^{-1} cm^{-2}, 400 cal s^{-1} cm^{-2}, and 35 cal s^{-1} cm^{-2}, respectively.

A simple bar graph of the thermal conductivity data would look like the one in Figure H-1. On the graph, the substances cork, concrete, copper, and zinc are numbered 1, 2, 3, and 4, respectively. Because the values for cork and concrete are small in comparison to those of zinc and copper, their bars are too small to see in this graph. Cork has the lowest thermal conductivity. Is it the best choice for your fireproof safe? At the end of this experiment you will prepare bar graphs similar to this one for several other parameters. Then you will use these graphs to help you make your final decision.

FIGURE H-1

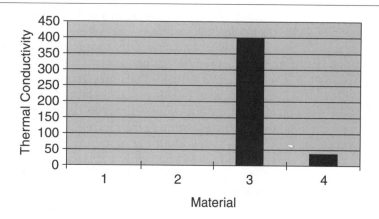

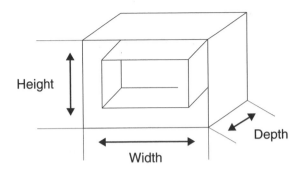

FIGURE H-2

A typical design for a fireproof safe is shown in Figure H-2 above. The inner box has these dimensions: height = 0.100 m, depth = 0.120 m, and width = 0.240m. The outer box has these dimensions: height = 0.134 m, depth = 0.154 m, and width = 0.274 m, making the spacing between adjacent surfaces 0.034 m. A material is to be selected for the interstitial space between the two metal boxes such that the following design specification is met.

Design specifications
1. The interior of the safe is not to exceed 177°C when exposed to a fire of temperature 954°C for half an hour.
2. The safe is to be portable such that an average person can carry it. Minimum weight is desired.
3. The safe is to be a consumer product. Cost is to be minimized.

Using an equation we will give you later, you can calculate the inner temperature of the safe after half an hour of exposure to the fire temperature for each interstitial material. The temperature must *remain less than 177°C* for the safe to be classified as fireproof. You can also calculate the mass of each material needed to fill the interstitial space in the safe. You can do this by calculating the volume between the inner and outer boxes and by measuring the density of each substance used in this lab. Then, mass equals the product of density and volume. For this calculation, we will assume that the geometry of the fireproof safe is fixed for all materials. You can determine the cost of each material needed to fill the interstitial space if you know the cost per unit mass. Finally, you will choose one of the materials you tested as best suited for the fireproof safe.

The specific heat capacity of a substance is determined by measuring the temperature change that results when the hot substance being studied is submerged in a measured amount of cooler water. In the first experiment, you determined the specific heat capacity for some metals that were heated by putting them inside a test tube heated in boiling water. This setup assumed that the metal and hot water were in close thermal contact and that, with time, the metal attained the same temperature as that measured for the hot water. This is generally a valid assumption, and it has the additional advantage of ensuring that the final temperature measured is due solely to the transfer of heat from the metal to the cool water and not from any drops of water from the hot water bath. By initially keeping the metal sample dry, no water from the hot water bath was transferred into the calorimeter beaker along with the metal.

This temperature assumption is considered valid because the pieces of metal were small enough to be in close physical contact with one another and with the sides of the test tube surrounded by the hot water. With the passage of 7 to 10 min it is reasonable to assume that the heat from the water bath has been transferred to all of the metal. What if it is not possible to obtain a sample that so conveniently fits inside a test tube? What do you do then?

In this experiment, you will measure the specific heat capacity of odd shaped pieces of zinc, of cork, and of concrete. Because test tubes are no longer appropriate containers for these solids, you must devise some alternate method that will allow close thermal contact between the solid and the hot water bath.

PROCEDURE

Part I: Formation of Groups

Form a group of three persons. Each person should be responsible for directing the experiment for one of the three substances. It is extremely important that each group member understand his or her role in today's experiment so that a reasonable conclusion can be drawn at the end.

Part II: Calorimeter Preparation

Determine the heat capacity of the calorimeter as in previous experiments.

Part III: Determination of the Specific Heat Capacity of Zinc (Group Work)

Measure the specific heat capacity of zinc using a modified method to keep the sample dry while it is being heated in the initial part of the experiment. Zip-lock plastic bags may be used if they are heat resistant. Manipulating the sample into and out of the plastic wrap or bag may require some practice. Is this something you can comfortably do alone or does the transfer of the solid from the plastic wrap or bag require help? You may want to practice the maneuver a few times before trying it at an elevated temperature. Allow time (10 min) for thermal equilibrium to be established. Record the temperature of the hot water bath in your notebook.

Repeat the measurement for a total of two trials. You will use an average of these two calculated values to answer the postlab questions.

Part IV: Determination of the Specific Heat Capacity of Cork and of Concrete (Group Work)

Repeat the procedure you worked out for the zinc measurement with samples of cork and of concrete. Do two trials for each substance. A different group member should set up and direct each of these. These substances require a full 15 min in the hot water bath for thermal equilibrium to be reached. Decreasing this time will result in very poor calculated values for specific heat capacity.

Part V: Density Determination

It is also necessary to determine the *density* of each substance. Density is one of the parameters included in the calculation of the inside temperature of a safe surrounded by a raging fire. Determine the densities of copper, zinc, cork, and concrete.

Individual Report

1. Calculate the specific heat capacity for all three substances (zinc, concrete, cork). Report the averaged value obtained from your two trials for each substance.
2. Prepare a bar graph containing the specific heat capacity values for copper, zinc, cork, and concrete. Label the substances on the graph. Which substance has the highest specific heat capacity? (If you considered only specific heat capacity values, you would choose the substance with the highest value at this point.)

3. Prepare a bar graph containing the density values of copper, zinc, cork, and concrete. Label the substances on the graph. Which substance has the greatest density? (If you considered only density values, you would choose the substance with the highest value of density at this point.)

4. Calculate the temperature inside a fireproof safe (T_{safe}) during a fire using each of the four candidates for filler material: zinc, copper, cork, and concrete. An estimation of T_{safe} can be obtained from the following.

$$T_{safe} = T_{fire} - (T_{fire} - T_{safe,initial})e^{\left(-\frac{hA_s}{c_s m}t\right)}$$

where $T_{fire} = 954°C$, $T_{safe,initial}$ = room temperature, and h is the heat transfer minus coefficient and is assumed constant for the three materials. (It is calculated to be 6 J (s^{-1} m^{-2} K^{-1})). A_s represents the exterior surface area of the safe in m^2, t represents time in seconds, and c_s and m refer to the specific heat and mass (in grams) of the interstitial material of the safe.

5. Prepare a bar graph of the inside safe temperatures for the four materials. Label the substances on the graph.

Group Report

Discuss the following and include the answers in the final report of each group member.

1. The equation used to calculate the inside safe temperature contains c_s as a variable. How does the material with the highest c_s value rate when the inside temperature is calculated? (Refer to your graph of heat capacity values for these substances.)

2. Do your experimental findings support the statement, "Substances with large specific heat capacity values are good 'heat storing' substances"? Why or why not?

3. Compare the ½-hour safe temperature calculated using each of the four materials. The inside temperature cannot exceed 177°C. The substances that meet the fireproof safe temperature specification are possible filler materials. List the possible filler materials.

4. Which property do you think is the most important in determining a good interstitial material for a fireproof safe? Why?

5. Prepare a bar graph containing the product of density and specific heat capacity for the substances you measured. On the graph, label the substances that had acceptable inside temperatures. Is there any correlation (agreement or lack of agreement) between inside temperature and the density-specific heat capacity product?

6. Which material—zinc, copper, cork, or concrete—is best suited as the filler material for the fireproof safe? Defend your choice. Did you include cost in your decision?

7. What happened to the thermal conductivity comparison that you made (i.e., how did that parameter enter into your choice for the interstitial material)? Did your intuition direct you to choose the lowest conductivity material (i.e., the cork)? Did your experiments and calculations support your intuition?

8. As a final check on your results, the next time you are in an office supply store, check the weight of a commercially available fireproof safe. Would you say that the manufacturer came to a similar conclusion as you did for the interstitial material of the safe?

EXPERIMENT GROUP I

CHEMICAL ENGINEERING PROCESS: MANIPULATING THE OUTCOME OF REACTIONS

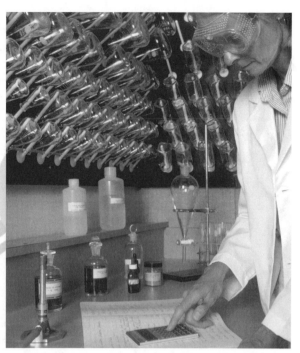

Calculating proper amounts of reactants is an essential part of many chemical syntheses.
Stephen Frisch/Stock, Boston/PNI

PURPOSE This group of experiments introduces students to chemical synthesis. Synthesis of a substance involves more than just mixing reactants in stoichiometric amounts. The order of addition of reactants and the methods used to isolate products can dramatically affect the outcomes of the reaction. When reactions are run continuously on an industrial scale, it is even important to control the direction of flow and how catalysts are used to speed up the reaction.

SCHEDULE OF THE LABS

EXPERIMENT 1: Skill Building Lab: "Synthesis of Metal Complexes"

1. Design a procedure to mix solutions of a transition metal salt and sodium saccharinate in appropriate amounts (group work).
2. Carry out the reaction of the substances (individual work).
3. Practice techniques of separation by filtration (group work).
4. Isolate saccharinate complexes (individual work).

EXPERIMENT 2: Foundation Lab: "Control of Reaction Products"

1. Design procedures to react a metal salt (group work).
2. Carry out the reactions with different isolation procedures (individual work).
3. Compare the results with the properties of known substances (group work).

EXPERIMENT 3: Application Lab: "Reaction Engineering"

1. Construct a tubular reactor (group work).
2. Load and prepare catalyst resin (group work).
3. Carry out a synthetic reaction with different reactor models (individual work).

SCENARIO Synthesis, or the construction of new chemical substances by a chemical reaction, is one of the simplest concepts in chemistry. We mix reactants in appropriate amounts, and then after a period of time we obtain a new compound. In actual practice chemical synthesis requires all the skills of a chemist, because controlling the outcome of a reaction depends on reactant ratios, temperature, and even the introduction of nonreactants to start, speed up, or direct a reaction.

In this group of experiments we introduce you to different parts of the "art" of synthesis. You will see how reactions are set up, how they are run, and how the products are obtained from the system. All of these steps are part of the design of the system, and this requires a knowledge of chemical reactions, an understanding of the safe handling of chemical substances, and a wide variety of tools for the efficient mixing, reaction, and separation of substances.

First, the usefulness of simple combination reactions will be shown as you and other students synthesize a set of chemical substances, related to each other but differing in the metal ion that they contain. The emphasis will be on the preparation of the reaction, and a critical pair of techniques, crystallization and filtration, will be included.

Second, you will see the way a reaction mixture is handled *after* the synthesis is done can affect the amount and even the composition of the products. This problem will be applied to a reaction that requires a separate isolation step. How isolation is done can convert a desired product into an undesired one.

The third reaction will cover the problem of catalytic reaction engineering. That is, what is the best way to contact a liquid feedstream with solid catalyst particles in order to maximize the amount of product (and profit, of course)? In a tubular reactor containing solid catalyst particles, the liquid flow can be directed either downward, which compresses the catalysts into a "packed bed," or upward, which gives rise to a "fluidized bed" if the liquid flows upward fast enough. The mode of operation has a big impact on the rate of reaction and the overall yield of product.

Experiment 1
SKILL BUILDING LAB: SYNTHESIS OF METAL COMPLEXES

Pre-Laboratory Assignment   **Due Before Lab Begins**

NAME: _____

Complete these exercises after reading the experiment but before coming to the laboratory to do it.

1. In this experiment, you will carry out a displacement reaction. A similar reaction occurs in the synthesis of complexes of metals with dimethylglyoxime (DMG):

   a. If you carry out this reaction with 2.00 g of $NiCl_2$, what mass of NaDMG will you need for the reaction?

   b. What is the theoretical yield of $Ni(DMG)_2$ you expect from the experiment in part (a)?

   c. If the actual reaction only gives 1.22 g of $Ni(DMG)_2$, what is the percentage yield of the reaction?

2. When copper(II) sulfate is prepared from copper(II) nitrate and sodium sulfate, the reaction initially produces a hydrated product:

$$Cu(NO_3)_2 + Na_2SO_4 + 5\ H_2O \rightarrow CuSO_4 \cdot 5\ H_2O + 2\ NaNO_3$$

   a. If you carry out this reaction with 2.00 g of $Cu(NO_3)_2$ and 1.00 g of Na_2SO_4, which reactant is the limiting reactant?

b. What mass of $CuSO_4 \cdot 5 H_2O$ do you expect from the experiment described in part (a)?

c. If $CuSO_4 \cdot 5 H_2O$ is heated, the water is removed, leaving anhydrous $CuSO_4$. What mass of anhydrous $CuSO_4$ will be produced from 2.20 g of $CuSO_4 \cdot 5 H_2O$?

3. Indicate the safety concern with saccharin. Is this also a problem with sodium saccharinate?

Experiment 1
SKILL BUILDING LAB: SYNTHESIS OF METAL COMPLEXES

BACKGROUND

The displacement reaction is among the most fundamental transformations in chemistry. In a displacement reaction, we substitute one atom or group of atoms with another. In its simplest scheme, this occurs with a *single* displacement reaction:

$$AX_n + nY \rightarrow AY_n + nX$$

A *double* displacement reaction can be viewed as an exchange of groups. This often happens with ionic compounds.

$$AX_n + nDY \rightarrow AY_n + nDX$$

Examples of double displacement reactions include

$$NaBr\ (aq) + AgNO_3\ (aq) \rightarrow AgBr\ (s) + NaNO_3\ (aq)$$

$$Na_2SO_4\ (aq) + BaCl_2\ (aq) \rightarrow BaSO_4\ (s) + 2\ NaCl\ (aq)$$

In the case of these reactions, we have a single product that is sparingly soluble in water. Therefore, the reaction is relatively simple: Mixing the solutions causes the immediate formation of a precipitate, and then that product can be isolated by filtering away the solution.

Other reactions, especially in organic chemistry and with most of the other transition metals, may occur in a similar fashion. But the rate of the reaction may be very slow. In those cases, there is plenty of time for other reactions to occur, giving undesirable side-products. We may need to heat the reaction mixture, to complete the reaction in a reasonable time.

For example, the formation of soap requires a simple displacement by hydroxide of three long-chain molecules, called fatty acids, from a molecule of glycerol:

$$
\begin{array}{l}
CH_2-O_2CC_{17}H_{33} \\
CH-O_2CC_{17}H_{33} \quad + 3\ NaOH \longrightarrow C_3H_8O_3 + 3\ C_{17}H_{33}CO_2Na \\
CH_2-O_2CC_{17}H_{33}
\end{array}
$$

This reaction can be done by cooking animal fat with very strong solutions of sodium hydroxide. Heat is needed because the rate of reaction of hydroxide in this case is slow.

The same problems can occur with transition metals. In this lab, you will see that the rate of addition of a molecule called saccharinate is slow at room temperature, but fast at higher temperatures. You will also find that the product $M(Sac)_2$ is much more soluble at elevated temperatures, so simple cooling is required for the isolation of the product.*

---

* This procedure is adapted from that described by Haider, S. Z., Malik, K. M. A., and Ahmed, K. J. 1985. *Inorganic Syntheses* 23:47–51.

$$MX_2 + 2\,Na-NaSac \longrightarrow 2\,NaX + MSac_2$$

NaSac

The saccharinate in this case is called a *ligand* for the metal. A ligand is a molecule or ion that forms a bond to a metal. Saccharinate is a ligand with a single negative charge, so two saccharinate ligands will complex with a metal with a +2 charge to form a neutral substance. Other ligands may be uncharged (e.g., water or ammonia).

An additional factor for saccharinate complexes is that they form with water molecules also coordinated to the metal. This means the material that forms is a *hydrate*, and the formula is properly written $MSac_2 \cdot 4\,H_2O$. The · symbol indicates that the water molecules are intact, but incorporated in the solid usually as neutral ligands to the metal. In many cases, hydrates can be dehydrated by gentle heating to give the anhydrous salt. Mass calculations for hydrates must be done with the waters included.

 Caution: This experiment includes the use of sodium saccharinate. Saccharinate is an anion formed by removing a hydrogen ion from saccharin, a substance that is still used today as an artificial sweetener. Both saccharin and saccharinate should be handled carefully because they have been implicated in animal studies as possible carcinogens.

The metal salts in this experiment have been chosen because they do not have significant identified toxicity. However, with all metal salts there are risks associated with high doses. Therefore, handle the solutions carefully.

PROCEDURE

Part I: Formation of Groups

You will be working in groups of three. Each student must design a procedure for a different metal saccharinate complex. The starting materials available are copper (II) sulfate, zinc nitrate, and iron(II) chloride.

Decide which student will work with which metal. You will all work with 1.00 g of sodium saccharinate and an appropriate amount of the metal. Carry out the stoichiometric calculation for the synthesis shown in the background.

In this laboratory, it is possible to work as if you were blindly following a recipe. This is wrong. Good laboratory technique requires that you keep careful notebook records of what you have done and what you observe in each step. Therefore, prepare your notebook in sections for the preparation of the reaction solution, processing of the reaction mixture to isolate the product, and collection of the product.

In the report for this experiment and the other experiments in this group, you must report what you observed at each stage. Leave space in your notebook to put observations about the reaction, including colors, whether the mixture is homogeneous or heterogeneous, and the characteristic appearance of the solids.

Part II: Design and Implementation of Synthetic Procedures (Group Work)

Use a centigram balance to measure out the correct amount of each reactant in separate small (25- to 50-mL) beakers. Dissolve each reactant in a small amount of water, between 10 and 20 mL each. The dissolution process may take several minutes.

Once both reactants are dissolved, combine them in one beaker and add a clean stir bar. Rinse the original beakers with 2 to 3 mL of deionized water. Place the reaction mixture on a stirrer hot plate, then start the stirrer. Stir gently to avoid spattering, then turn the heat on a moderate (about halfway on the scale) setting.

Monitor the reaction every few minutes to be certain that there is water left and no large amount of solid has formed. The reaction will come to boiling. Once you see solid formation around the edges of the solution, remove the beaker from the hot plate and leave it to cool until it is warm to the touch. At that point, you may put it in an ice bath.

Part III: Practice in Vacuum Filtration (Individual Work)

You will separate the product of the reaction by crystallization and filtration through a Büchner funnel. This is a flat ceramic or plastic device that has many small holes. The holes are too large to keep back most substances. So, a piece of filter paper is placed atop the holes.

The filtration is done by putting the funnel on top of a *filter flask,* an Erlenmeyer flask with a side arm near the top. This side arm is hooked up to a vacuum source, usually a vacuum aspirator. This creates a vacuum suction by the rapid flow of water through a small tube. You connect the filter flask to the aspirator using a piece of rubber tubing.

The filtration apparatus must be carefully secured with clamps because it is top-heavy. Also, be careful when removing the vacuum. Do *not* just turn off the aspirator water; this may create a disastrous back-flow of the water into the filter flask. Instead, break the vacuum by disconnecting the rubber tubing from the filter flask.

Begin with a separation of a sample of the compounds salicylic acid and sodium chloride. You will be given a mixture of the two. Obtain about 2 g of the mixture in a small beaker (record the exact mass to ±0.01 g).

The sodium chloride is very soluble in water; the salicylic acid is not. Add about 10. mL of water to the mixture. Swirl gently. Turn on the vacuum, then pour the heterogeneous mixture into the filter flask. If any solid comes through the filter paper, then you have not set up the system correctly; redo the experiment. If the solution that comes through (called the filtrate) is clear, then you have done the filtration correctly. Use a wash bottle to rinse any remaining solids from the beaker into the Büchner funnel. Then rinse the solid sitting on the filter paper with a small amount of water.

The flow of air through the filter funnel should remove most of the water from the solid. You may gently stir the filter "cake" with a spatula to aid drying.

Once you are certain that the material is dry, break the vacuum and then collect all of the solids in a clean, dry, and *pre-weighed* beaker. Obtain the mass of salicylic acid, then use this to calculate the mass of salicylic acid in the mixture.

Part IV: Crystallization and Separation of Product (Individual Work)

After practicing with the filtration, you should be ready to isolate the saccharinate salt. If immersion in ice water has *not* produced a significant amount of crystals, then ask your instructor for permission to add seed crystals to the reaction solution. Add *at most* three crystals of the target compound, then wait another 15 min.

The isolation is done by separation with the Büchner funnel. In this case, chill the wash bottle in an ice bath before use. Use this ice-cold water for all rinses.

Dry the product on the Büchner funnel, then collect it and weigh it.

REPORT

One part of your report should indicate the mass of salicylic acid you recovered from the mixture of salicylic acid and sodium chloride.

In your report, include the procedure that you used to prepare the metal complex. This should be written in the standard form of a synthetic chemical report. Write the paragraph in the past tense and in the *passive* voice. For example, write "A 2.00 g sample of the compound was weighed" not "I weighed a 2.00 g sample of the compound." Include the mass amounts of all reagents, the volumes of liquids, and the mass amount of the final product. Note the colors observed as well.

Tabulate:

a. The mass of product possible from the mass of metal salt used.
b. The mass of product possible from the mass of saccharinate used.
c. The mass of product collected in the lab.
d. The percent yield.

Use (a) and (b) from your table to demonstrate that you did mix the two reactants in the correct stoichiometric amount.

For your conclusion include a comparison of the properties and yields of products obtained by all the members of your group. Properties include the appearance, and also observations about the formation of the materials.

Experiment 2
FOUNDATION LAB: CONTROL OF REACTION PRODUCTS

Pre-Laboratory Assignment   **Due Before Lab Begins**

NAME: _____

Complete these exercises after reading the experiment but before coming to the laboratory to do it.

1. Cobalt forms complexes with the molecular ion oxalate ($C_2O_4{}^{2-}$) and ammonia. Sodium is present as a cation for the solids that are isolated, $Na[Co(C_2O_4)_2(NH_3)_2]$ and $Na_3[Co(C_2O_4)_3]$:

   a. Determine the oxidation number of Co in these complexes.

   b. A skeleton for this reaction involves a source of cobalt, sodium oxalate, and ammonia:

   $$\underline{}Co(NO_3)_3 + \underline{}Na_2C_2O_4 + \underline{}NH_3 \rightarrow$$

      i. Determine the balanced chemical reaction required to form $Na[Co(C_2O_4)_2(NH_3)_2]$. Include any other products.

      ii. Determine the balanced chemical reaction required to form $Na_3[Co(C_2O_4)_3]$. Include any other products.

2. In one part of this experiment you have to add hydrogen peroxide to a solution containing Co(II) ion to produced Co(III) ion. A solution contains 0.40 moles per liter hydrogen peroxide. How much of this is needed to completely react with Co(II) from 1.00 g of $CoCl_2$?

3. In the skill building lab, you were told that the transition metals iron, copper, and zinc do not have significant toxicity associated with them. Is this true for cobalt?

Experiment 2
FOUNDATION LAB: CONTROL OF REACTION PRODUCTS

BACKGROUND

In the Skill Building experiment you worked on a reaction that could be done in the same way with different metals. The products reflected the nature of the metal in each, through their color and crystalline appearance.

In this week's experiment, you will investigate a different kind of synthetic variation, associated with changes in the process used to isolate a product. You'll also see an important restriction that can be important in the way that reactions are done.

The synthetic goal of these reactions are complexes of Co(III) with the nitrogen-containing ligand ethylenediamine. Some of these are

$[Co(en)Cl_4]^-$
blue

cis-$[Co(en)_2Cl_2]^+$
green

$[Co(en)_3]^{3+}$
yellow-orange

In carrying out experiments to selectively make the complexes with two and three ethylenediamine ligands, a chemist would use known facts about Co(III) complexes and their syntheses. You should remember these, too.

1. Your work last week suggests one route to making these complexes: mixing a simple metal salt such as $CoCl_3$ with ethylenediamine. However, it is known that Co(III) is very slow to react in displacement or addition reactions. Therefore, you will start with Co(II) salts. These will be oxidized by the addition of hydrogen peroxide to convert the Co(II) to Co(III):

$$2\ Co^{2+}\ (aq) + 2\ H^+\ (aq) + H_2O_2\ (aq) \rightarrow 2\ H_2O\ (l) + 2\ Co^{3+}\ (aq)$$

2. In the absence of an excess of added chloride, these complexes are all a red-brown color. The distinctive colors indicated earlier require the presence of a large excess of hydrochloric acid. In fact, *concentrated* hydrochloric acid is needed.

3. It is impossible to "simply" assemble the complexes by addition of a calculated amount of ethylenediamine. Different amounts of the ethylenediamine may give different chemical products.

Caution: This laboratory has several components that present hazards. All can be managed well with attention to personal protection through gloves and proper procedures. In particular, the use of chemically resistant gloves is suggested. And, as always, be certain to wear proper eye protection and to thoroughly wash your hands at the end of the laboratory.

You will be working with ethylenediamine and concentrated hydrochloric acid. Neither is listed as a toxin, but they are corrosive if handled incorrectly. The HCl solutions typically emit a vapor of HCl gas that can be very irritating. Work with this *only* in a properly ventilated environment, and be very careful not to breath the fumes over the solutions. The cobalt(II) salts used in this experiment are possible carcinogens if ingested. There is no airborne hazard, but spills should be cleaned up immediately.

Part I: Formation of Groups

You will be working in pairs, sharing the basic reaction mixture but individually carrying out different isolation steps. The starting materials available are solid cobalt(II) chloride and solutions of ethylenediamine and hydrogen peroxide. Concentrated hydrochloric acid will be available in the fume hoods. Different groups will be assigned different amounts of ethylenediamine.

Part II: Preparation of a Reaction Solution (Group Work)

Make careful notes of the changes in color that occur at different points. When groups compare their results, you may find significant differences even *before* the final isolation of the product.

The reaction will be carried out with 2.00 g of cobalt(II) chloride hydrate. Take the water into consideration when calculating the number of moles you are using.

Weigh the salt on a centigram balance in a clean, dry 250-mL beaker. Add a small amount of water to dissolve.

Groups will be assigned by the instructor to add 2, 3, or 4 mol of ethylenediamine per mole of cobalt. Calculate the actual number of moles of ethylenediamine you will need, then determine the volume of ethylenediamine solution you require. Measure the ethylenediamine solution using a graduated cylinder. Mix the reaction solution well, then wait 10 min before proceeding.

The hydrogen peroxide should be added last. Calculate the actual number of moles of hydrogen peroxide you will need based on the number of moles of *cobalt*. Measure the hydrogen peroxide solution using a graduated cylinder. Mix the reaction solution well, then wait 10 min before proceeding.

Part III: Processing the Product (Individual Work)

Each student will take half the reaction mixture and process it to isolate the product. The two "processing" methods you should use are

1. Heat half of the reaction solution in the fume hood to concentrate it to about 10 mL, then cool the solution and add concentrated HCl to neutralize the solution. Once the solution is neutralized, add 10 mL of excess concentrated HCl.
2. Neutralize half of the reaction mixture with HCl before heating, then reduce the volume with heating to about 10 mL. After cooling, add 10 mL of excess concentrated HCl.

Part IV: Isolating the Product (Individual Work)

If immersion in ice water after processing has *not* produced a significant amount of crystals, then ask your instructor for permission to add seed crystals to the reaction solution. Since you do not know what product will be formed from which method, you must add a small amount of a mixture of the possible products, which will be available from the instructor. Wait another 15 min before collecting any solids.

The isolation is done by separation with the Büchner funnel. Use 95% ethanol solution in water for all rinses.

Dry the product on the Büchner funnel, then collect it and weigh it.

REPORT

In your report, include the procedure that you used to prepare the metal complex. This should be written in the standard form of a synthetic chemical report. Write the paragraph in the past tense and in the *passive* voice. For example, write "A 2.00 g sample of the compound was weighed" not "I weighed a 2.00 g sample of the compound." Include the mass amounts of all reagents, the volumes of liquids, and the mass amount of the final product. Note the colors observed as well.

Show stoichiometric calculations for the possible products. Tabulate

a. The masses of the products $[Co(en)_3]Cl_3$ and $[Co(en)_2Cl_2]Cl$ possible from the mass of metal salt used.

b. The masses of the products $[Co(en)_3]Cl_3$ and $[Co(en)_2Cl_2]Cl$ possible from the number of moles of ethylenediamine used.

c. The mass and the major component (as indicated by color) of the product you actually got.

d. The processing method you used.

e. The percent yield.

Compare the mass and major component of the product you obtained with those of your partner. Comment on the results of the two processing methods.

Tabulate class data and compare masses and the major component of the product obtained by groups using the same and different numbers of moles of ethylenediamine initially. Account for differences.

Experiment 3
APPLICATION LAB: ENGINEERING A REACTION

Pre-Laboratory Assignment   **Due Before Lab Begins**

NAME: _____

Complete these exercises after reading the experiment but before coming to the laboratory to do it.

1. Compare and contrast the terms "synthesis" and "decomposition."

2. Give a balanced chemical equation for the reaction you will study today.

3. Salicylic acid can be converted into acetylsalicylic acid ($C_9H_8O_4$), a compound commonly known as aspirin:

$$C_7H_6O_3 + C_4H_6O_3 \rightarrow C_9H_8O_4 + C_2H_4O_2$$

   A student has 100 mL of solution containing 1.00 g of salicylic acid as the limiting reactant. What mass of aspirin should result?

4. What is the problem in the separation of products from reactants in this laboratory?

5. Indicate why open flames are unacceptable during this procedure.

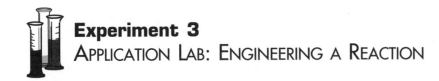

Experiment 3
APPLICATION LAB: ENGINEERING A REACTION

Many chemical reactions, like the ones you may have used during the last two weeks, can be done in an hour or less, depending on the temperature and concentration of the reactants. But, it may not be enough to "mix and wait" for a reaction. In many cases, another chemical substance, called a catalyst, is needed.

A catalyst is *not* part of the reactants in a chemical equation; nor is it one of the products. It may be essential in giving the reactants a chance to make the products. This is because the reactants may not react fast enough, or with the right outcome, on their own.

Catalysts are found in many different chemical systems. The enzymes that are essential for all life are catalysts. Virtually all cars have a catalytic converter to consume any fuel that is not burned in the engine and to break down unwanted products of combustion before they are emitted into the atmosphere. Synthetic chemists also use catalysts. Some catalysts approach the complexity of enzymes; others are simple acids and bases that change the properties of aqueous solutions.

In this experiment, you will use a catalyst to promote a reaction that breaks a carbon–oxygen bond. The chemistry of carbon-containing compounds, known as organic chemistry, is important in biological systems, but it also forms the basis of the petrochemical, pharmaceutical, and fine chemical industries. The incredible variety of carbon-containing compounds hides two important facts. First, bonds to carbon can be very slow to form, because we must often break a bond to a carbon atom first. Second, breaking carbon–oxygen bonds takes a lot of careful control, because a reaction to make one bond can easily give rise to many different products.

In the previous experiments you studied synthetic reactions in which two or more components were combined to give new compounds. In this experiment, you will study decomposition reactions. In particular, we will break a C—O bond in an *ester* with water, a reaction known as a hydrolysis (literally, "splitting by water") reaction.

In this reaction, water splits methyl salicylate, $C_8H_8O_3$, into salicylic acid, $C_7H_6O_3$, and methanol, CH_3OH.

Engineering with a Solid Catalyst

Many different reactions use a catalyst that is dissolved in a solution with the reactants. This is easy to do, but it has the disadvantage that the catalyst may be hard to remove from the product mixture. Some catalysts are very expensive, so it is important that they be in a form that can be used in many different reactions. For this reason, chemists and engineers have paid attention to the possibility of making catalysts into easy-to-separate solids. This is true, for example, of the solids in an automobile's catalytic converter.

In this reaction, you will use a catalyst that is made with small beads of polystyrene, known as a resin. The resin supports, or anchors, the catalyst. You will be asked to construct a catalytic reactor to start the experiment. This will be built from a simple glass tube with connections on the top and bottom.

 Caution: The ester under study is called methyl salicylate and, as you will soon understand, it has the nickname oil of wintergreen. Very small amounts of this are found in foods, but higher amounts can be harmful, even fatal, if ingested. The product salicylic acid lacks the odor of the ester, but it is still active as a relative of aspirin. Normal safety precautions will provide ample protection.

The reaction is run with a solvent that is 95% ethanol. This can be irritating to the skin. It is also flammable. *No open flames or other ignition sources are allowed in the lab with this procedure.*

The NaOH and HCl that are used in the final separation steps are caustic and corrosive. Be careful to avoid contact with the skin.

PROCEDURE

Part I: Formation of Groups

Students will work in groups of two. Each group will build a catalytic reactor to share, and then each student will use the reactor to test different ways to engineer a chemical reaction. Obtain a column, a syringe barrel to serve as a reservoir, and the necessary tubing.

The key to building the apparatus are three-way valves, called stopcocks, that allow you to control the flow of liquid into and through the reactor. The valves have a threaded connection (an "outer" joint) that can accept the other "inner" joint with a gentle twist. Either joint can also be connected to ⅛-in. tubing that serves to connect the pieces.

Start by inspecting the three-way valve and experimenting with it and some water. Make sure you understand that the handle on the valve indicates which part of the system will *not* have fluid flowing through it (Figure I-1).

Part II: Preparation of the Tubular Reactor with Resin (Group Work)

Measure 3 g of resin into a 100-mL beaker. Record the mass of the resin to the nearest 0.01 g. Add 10 mL of deionized water to the beaker. If necessary, use a pipet to rinse any resin that sticks to the side of the beaker down into the bottom of the beaker. The solid resin and the water form a heterogeneous mixture called a *slurry*.

All column attachments are shown in Figure I-2. Attach a three-way valve to the bottom of the column (use the "inner" joint that is across from the "outer" joint of the valve to make this connection). Secure about 12 in. of tubing to the outer joint of each stopcock. Add water to the column, then open the stopcock and drain the water from the reactor. Add more water to half-fill the column.

FIGURE I-1 HANDLE POSITIONS FOR THE THREE-WAY VALVE

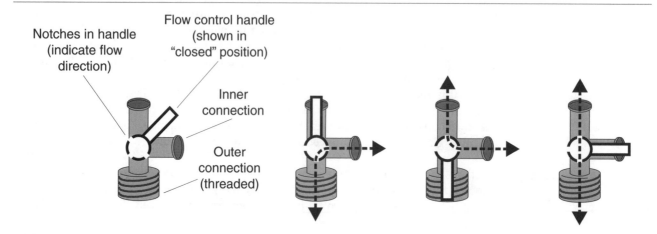

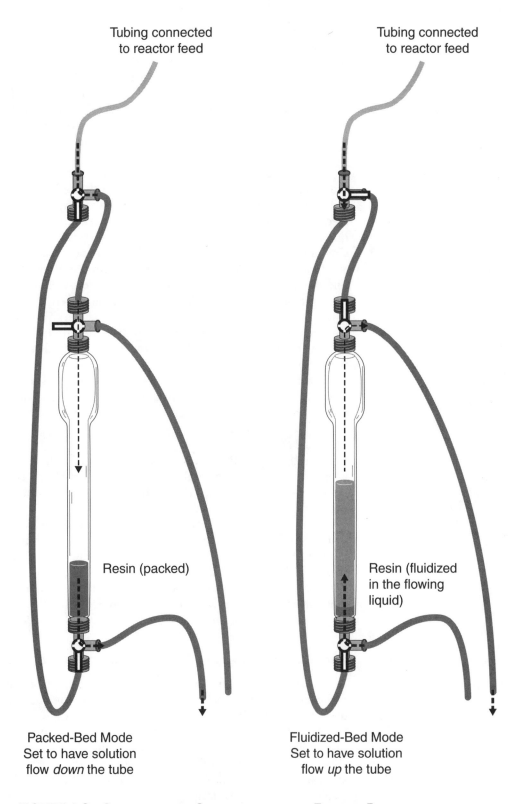

Tubing connected
to reactor feed

Tubing connected
to reactor feed

Resin (packed)

Resin (fluidized
in the flowing
liquid)

Packed-Bed Mode
Set to have solution
flow *down* the tube

Fluidized-Bed Mode
Set to have solution
flow *up* the tube

FIGURE I-2 SCHEMATIC FOR CONSTRUCTION OF A TUBULAR REACTOR

To put the slurry into the column, you will need to gently swirl the beaker, then pour the contents of the beaker into the top of the column. You should drain the column back into the beaker to get liquid to rinse the beaker. Continue until all the resin is transferred into the column, then fill the column with water. Put another three-way valve on the top and attach two 12 in. lengths of tubing to the valve.

Complete the assembly by securing the column with a clamp. On a separate stand, clamp a syringe or other reservoir as high as possible above the column. Use separate pieces of tubing to connect the valve on the syringe to the top and bottom of the column.

You will use this reactor with a solution of methyl salicylate. The liquid leaving the reactor is called the effluent. Note that the resin is held in place by a glass disk at the bottom of the tube. There are then two ways of contacting the flowing liquid feedstream with the catalyst beads in the reactor. First, the liquid can be directed downward through the reactor, which will push the resin beads against the disk. Chemical engineers refer to this mode of operation as a *packed bed*.

In a second mode of operation, flow can be directed upward through the reactor. (Since the outlet where the effluent is collected is well below the level of the feed tank, gravity will pull the liquid through the system, even if sometimes, as in the reactor, the flow has to go upward for a while.) In this configuration, the resin particles are unconstrained by a glass disk and are buffeted about in becoming a *fluidized bed*. In this situation the upward push by the fluid is eventually balanced by the weight of the particles.

Practice with both modes by using water in the syringe. You can adjust the size of the valve openings to control the rate of flow. You should use a graduated cylinder to catch the outflow. Practice getting the flow rate to be the same in both modes at about 20 mL per minute.

Once you are confident of your flow settings, you are ready to activate the resin. Remove the top of the column. Drain the solution from the column until the liquid is just at the top of the catalyst. Then, add 30 mL of 0.10 M (mole per liter) HCl to the column. Drain the liquid until the top of the liquid is again just above the top of the catalyst bed. Wait 10 min.

You need to remove any free HCl from the column by rinsing well. Add deionized water to the column and allow the column to drain. Continue adding water and draining until the liquid that comes off the column is in the neutral range for pH (use pH paper that shows a full 2–12 range; you are looking for a value between pH 5 and 7).

The reactor now has to be treated with 95% ethanol (*not* containing any reactants) to coat the resin with the correct solvent mixture. Run a total of 60 mL of 95% ethanol down the column. Leave the column full of 95% ethanol; replace the cap.

Part III: Preparing the Reaction (Group Work)

Ask your instructor for the solution of methyl salicylate in ethanol. Each student will determine a reaction yield; one student will use the packed bed mode, the other will use a fluidized bed mode. Assist each other as you do your experiment.

After ensuring that all valves are in the closed position, connect the syringe or reservoir containing the feed solution to the reactor setup. Open the valves successively from the syringe or reservoir to the reactor inlet, opening the last valve slowly and carefully so that the liquid flow rate can be controlled. Catch the exit stream in a 250-mL Erlenmeyer flask.

Once air bubbles are cleared from the circuit, direct the effluent to a graduated cylinder, and using a stopwatch, adjust the valve such that the flow is 20 mL/min. Once this flow rate has been established, wait another 3 or 4 min while the reaction attains "steady state," that is, a state of smooth, continuous operation with no changes in any system variable. (Recall, for example, that since the reactor is initially filled with 95% ethanol, it will take a little while for all of the ethanol to be replaced by the feedstock.) After you feel that steady state has been attained, collect 100 mL of effluent. Record the time it takes to do this, in order to calculate the exact flow rates). The effluent contains both unreacted methyl salicylate and the product salicylic acid. Separate the product from the reactant as described in Part IV.

A fluidized bed can be created by switching the appropriate valves to achieve upward flow in the reactor. Adjust the valves such that the flow rate is the same as before, 20 mL/min. Again, wait a few minutes until steady state is obtained in the fluidized bed mode, and then collect 100 mL of effluent, and separate reactant and product as before. In the report, you will compare the performance of these two modes.

Part IV: Processing the Reactions

In a fume hood, gently heat the beaker containing your sample to reduce the volume to about 5 to 10 mL. As this happens, the ethanol evaporates faster then the water, so that the liquid that remains is close to 95% water. If a solid appears, add 95% ethanol dropwise to make the system homogeneous.

The isolation of the salicylic acid is complicated by the presence of the unreacted methyl salicylate. Though methyl salicylate is a liquid, not a solid, it can adhere to and mix with the salicylic acid and cause the formation of an oily or gummy product. We solve this by adding just enough 1 M NaOH to selectively dissolve the salicylic acid in water. Then we remove the liquid methyl salicylate that separates out.

Take the reaction product and add NaOH in small portions (about 0.50 mL at a time) until the mixture appears clear. There may be separate drops of methyl salicylate at the bottom of the beaker, but they should be clear also.

Weigh a small beaker or flask. Use a pipet to transfer the methyl salicylate to the beaker or flask, and obtain the weight again. This allows us to calculate the weight of the "recovered" methyl salicylate.

The salicylic acid has been converted to a base by the NaOH, and this is water soluble. Crystallization can be done by reconverting it to the acid form. This is done by adding 1 M HCl to the solution until the solution is just acidic, by pH paper test. Solid salicylic acid may appear. Set the mixture aside for 10 min to complete the separation.

Collect any solids by vacuum filtration. Rinse the solids with some water, then collect them on a watch glass. When they are completely dry, weigh them.

REPORT

For both reactor modes, calculate the yield from both the recovery of the left over reactant, methyl salicylate, and the product, salicylic acid. For each mode, how well do the two yield measurements agree? What might be the source of error in each case?

In the packed bed mode the reaction begins at a relatively high rate as soon as the reactants hit the catalyst bed. Farther down the reactor, the concentration of methyl salicylate decreases and the concentrations of products increase. A chemical engineer can model the concentration profiles along the length of the reactor, but for the present purposes this isn't necessary.

In the fluidized bed, the concentration of reactants in the liquid phase immediately drops when the feed enters the reactor because the catalyst bed is well mixed. This means that the concentration of reactant in the fluidized bed is on average lower than the concentration in the packed bed. This is the key difference in operating modes. Explain if your yield measurements for the two cases are consistent with this difference.

EXPERIMENT GROUP J

ECOLOGICAL ELEMENT CYCLES: JUDGE LEVELS OF ENVIRONMENTAL DISTURBANCE

The application of nitrogen is common on many farms.
Larsh Bristol/Visuals Unlimited.

PURPOSE   In this experiment group, you will utilize three different analytical methods—gravimetric, instrumental, and titrimetric—to perform elemental analyses of two different ecologically important elements, nitrogen and phosphorus. Determining elemental composition is not just an academic activity. We need to keep track of the amount of elements if we are to study the movement of elements through the ecosystem.

- *Gravimetric analyses.* The analyte (the substance to be determined) is physically separated from all other components of the sample and the solvent. Precipitation is a common method used to chemically separate the analyte. Weight is the only measurement needed in gravimetric analyses.
- *Titrimetric methods.* These involve measuring the volume of a solution of known concentration needed to react with the analyte or some derivative of the analyte.
- *Instrumental techniques.* An instrument is used to measure the quantity of analyte present in the sample. The most common instrumental techniques utilize spectrophotometry (such as UV–visible, infrared, and atomic absorption) or chromatography, such as gas chromatography (GC) and high-performance liquid chromatography (HPLC).

The method that a scientist uses depends on several factors. The most important are accuracy, cost, and convenience. Instrumental methods may be more expensive in the short run, but they may be much easier and more reliable. On the other hand, so-called wet methods, such as gravimetric and titrimetric methods, are often less expensive and easier to do in the field. The nature of the analyte also affects this decision. Gravimetric methods need an analyte in large enough amounts to be weighed in some form. If a physical separation is needed, a gravimetric method may be the simplest one to use. In a titration, one needs a method to measure the change in solution composition that signals the end point. And in spectrophotometry the analyte must be completely converted into some species that absorbs light without interference.

SCHEDULE OF THE LABS

EXPERIMENT 1: Skill Building Lab: "Gravimetric Analysis of Phosphorus"

1. Determine which metal cation will selectively precipitate the phosphate ion (group work).
2. Determine the percent by mass of phosphorous in an unknown mixture (individual work).
3. Set up the NO_2 collection tubes for the second week of the experiment group (individual work).

EXPERIMENT 2: Foundation Lab: "NO₂ Analysis"

1. Prepare a calibration curve using standard solutions (group work).
2. Determine the concentration of NO_2^- in a prepared unknown (individual work).
3. Determine the amount of NO_2 collected in the tubes (individual work).
4. Set up the soil analysis experiment to be completed the third week of the experiment group (individual work).

EXPERIMENT 3: Application Lab: "Nitrogen and Mud"

1. Determine the percent by mass of mineral nitrogen in disturbed and undisturbed soil samples (individual work).
2. Determine the mineral nitrogen content of a standard solution (individual work).
3. Dry a soil sample (individual work).
4. Organize and analyze group and class results (group work).

SCENARIO

Suppose your family decided to build a cabin, a place to go to in summer, up north, in the woods. "Will building the cabin disturb the environment?" your mom wonders. Your dad, remembering the cost of tuition, says "From what she learned in college, she should be able to tell us."

You remember some techniques you learned in introductory chemistry that would be helpful. You explain that it is important to measure the response of two key elements, phosphorus and nitrogen, that all organisms need for growth and reproduction. Nitrogen is needed to construct amino acids and the enzyme that plants use to fix carbon. Phosphorus is needed for important molecules such as DNA and ATP.

"What exactly do we measure?" your mom asks. You explain by drawing the nitrogen cycle for the site in the woods where the cabin will be built (Figure J-1). You point out the places in the cycle where nitrogen is in an inorganic form—NH_4^+, NO_3^-, and NO_2. There are two reasons to focus on these compounds, you explain. First, inorganic compounds should be easy to measure chemically because they are not part of complex organic molecules. Second, concentrations of inorganic nitrogen and phosphorus are normally low, because bacteria, fungi, and roots compete for each atom of nitrogen and phosphorus.

"How do we go about making the measurements?" asks your dad. You remember a diffusion method for measuring NH_4^+ and NO_3^- (known as mineral nitrogen) in small soil or water samples. You also remember a passive method for sampling NO_2 in air.

What you can't remember is what indicates environmental disturbance, higher or lower NH_4^+ and NO_3^- concentrations in disturbed relative to undisturbed soil? And is environmental disturbance indicated by higher or lower NO_2 concentrations in air?

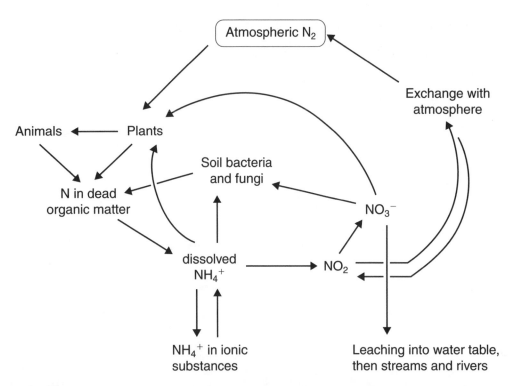

FIGURE J-1 A TERRESTRIAL NITROGEN CYCLE EMPHASIZING PORTIONS THAT OCCUR IN THE SOIL

Experiment 1
SKILL BUILDING LAB: GRAVIMETRIC ANALYSIS OF PHOSPHORUS

Pre-Laboratory Assignment   **Due Before Lab Begins**

NAME: _____

Complete these exercises after reading the experiment but before coming to the laboratory to do it.

1. What volume of 0.25 M $CaCl_2$ is needed to completely react with 0.576 g of Na_3PO_4?

2. Calculate the mass of phosphorus in 1.454 g of $Sr_3(PO_4)_2$.

3. Calculate the percent by mass of P in P_2O_5.

4. a. What is the mass of P_2O_5 that contains 0.164 g of P?

   b. If a 2.500-g sample contains 0.164 g of P by mass, what is the percentage of P_2O_5 in the sample?

5. The phosphorus content in a fertilizer is reported as 23% P_2O_5 by mass. What is the mass of phosphorus in a 1.575-g sample of the fertilizer?

6. One of the solutions used in this experiment may contain a lead compound. While you are expected to handle all chemicals carefully, why must you be particularly cautious with a solution containing a lead compound?

Experiment 1
SKILL BUILDING LAB: GRAVIMETRIC ANALYSIS OF PHOSPHORUS

BACKGROUND

Phosphorus is an essential element for plant growth. The phosphorus must be part of a water soluble compound to be useful to plants. Plants usually absorb phosphorus as $H_2PO_4^-$. Although the phosphate ion, PO_4^{3-}, is common in soil, it is often present in an insoluble salt such as $Ca_3(PO_4)_2$.

People often apply fertilizers to their soil to increase the concentration of phosphorus as well as nitrogen and potassium, two other essential minerals. A common fertilizer is labeled 5-10-5. The numbers represent the percentages, by mass, of nitrogen, phosphorus, and potassium, respectively, in the fertilizer. The phosphorus is reported as a percent by mass of P_2O_5 in the fertilizer, even though the fertilizer does not contain any P_2O_5. The phosphorus is generally present as $(NH_4)_2HPO_4$ (so this compound can also serve as a source of nitrogen for the plants).

In this experiment you will determine whether a mixture contains an appropriate percentage of phosphorus for use as a fertilizer. Fertilizers generally contain between 10 and 27% P_2O_5 by mass. Therefore, you will perform a gravimetric analysis to determine the percent by mass of phosphorus in your unknown mixture.

The measurement involved in gravimetric analysis is weight, so it is important to weigh as accurately as possible. You should use an analytical balance throughout this experiment. When gravimetric analyses are used to chemically separate an analyte, the process is very time-consuming because great care is taken to obtain a pure and completely dry final product. These procedures are only applicable when high concentrations of analyte are present. Your samples all contain enough phosphorus so that the gravimetric technique is appropriate in this context.

 Caution: One of the test solutions used in this experiment may contain a lead compound. Lead compounds are *poisonous*. Handle with care. Dispose of all lead-containing compounds and solutions in a designated waste receptable. Some of the solutions are caustic (have a high pH), and these can be very bad for the eyes and skin.

PROCEDURE

Your unknown consists of a mixture containing the analyte (phosphorus—as the phosphate ion) and other substances, some of which are water soluble and some of which are insoluble. Your goal is to determine the percent by mass of phosphorus (as P_2O_5) in the mixture. You first need to determine what metal cation will selectively precipitate the phosphate ion. You will then precipitate the phosphate with the appropriate reagent, collect, dry, and weigh the precipitate.

Part I: Formation of Groups

You will be working in a group of three to four people. Everyone is expected to participate in the experiment, to record data in his or her own notebook, and to complete his or her own report.

Each group should obtain a bottle of a sample with an unknown amount of phosphorus. The unknown will contain 10 to 15 g of sample. Divide the sample among group members. Be sure to save some of the sample in case anyone in the group encounters a problem during the experiment and needs additional sample to complete the analysis. Each person in the group will individually analyze a different sample of the same unknown.

During the lab you will need to heat a mixture on a hot plate or a burner. Check with your instructor. If a hot plate will be used, get one for the group and set the temperature to a low setting (about 25%).

Part II: Selective Precipitation of the Phosphate Ion (Group Work)

Your unknown may contain several different anions in addition to the phosphate ion. Before you analyze your unknown you need to determine which metal cation will precipitate the *phosphate* ion only. To accomplish this task, you will test a series of sodium salts, each containing a different anion, with various metal cation solutions. You will use spot plates for your tests, so you only need 10 to 20 drops of each of the solutions per test. Record all observations.

Part III: Determination of the Percent by Mass of Phosphorus in a Mixture (Individual Work)

The analyte (the phosphate ion) in the sample is present in a water soluble compound. Other components of the mixture include water insoluble substances. As a group, design a way to separate the water soluble components from the insoluble components. Have your lab instructor approve your procedure before you proceed. Once your group has decided upon a procedure, you will each work independently on your own samples.

Since weight is the critical measurement in this analysis, weigh approximately 2 to 3 g of sample on an analytical balance. Record the exact mass.

First you need to separate the water soluble components from the insoluble components. Using the technique designed by your group, separate the water soluble components from the mixture. *Save the insoluble components* for weighing when dry in order to determine the percent by mass of the insoluble components in the original mixture.

The phosphorus (in the phosphate ion) is present in the soluble component of the mixture. You will precipitate the phosphorus out as an insoluble phosphate. To do so, assume the original sample is entirely Na_3PO_4 and calculate the volume of the metal cation solution (chosen in Part II) needed to completely react with this amount of Na_3PO_4. Check the label on the bottle of solution of the metal cation for the molarity of the solution.

Add the calculated volume of the metal cation solution to the water soluble component of your mixture.

Digesting the Precipitate

Your precipitate at this point will probably be a very fluffy solid suspended in a solution. Though the solid is denser than the solution, the small particle size of an initial precipitate usually prevents good settling. To separate the precipitate well, you need to warm the mixture. This process is called digestion and is necessary to enlarge the particle size of the precipitate.

Bring the solution to a gentle boil and then lower the flame or decrease the power on the hot plate. Cover the beaker with a watch glass and maintain a low heat for about 30 min. Be careful not to boil the solution again. Allow the precipitate to settle; it should be denser and more crystalline now.

Label and then weigh a watch glass and one piece of filter paper. Record this mass. Place the filter paper in a funnel. While the solution is still hot, quantitatively transfer the precipitate to the funnel. Remove any precipitate on the beaker's walls by using a rubber policeman and deionized water. Pour all rinse water onto the filter paper. Wash any precipitate clinging to the rubber policeman onto the filter paper with water. Finally, rinse the precipitate on the filter paper with water to remove any unwanted residual solution from the precipitate.

After the water has stopped dripping through the funnel, test the filtrate for any unreacted phosphate by adding several drops of the metal cation solution used to precipitate the phosphate. Record your observations. When you are certain the filtrate has no residual phosphate, carefully remove the filter paper and precipitate from the funnel. Transfer the filter paper and precipitate onto the watch glass and allow the precipitate to dry.

When they are completely dry (next lab period), weigh the watch glass, filter paper and precipitate.

Part IV: Preparation of Nitrogen Dioxide Collection Tubes (for Week 2)

During the second experiment of this group, you will determine the atmospheric concentration of nitrogen dioxide at different sites. See Experiment 2 in this group for details regarding passive diffusion gas collection and the safety precautions for the experiment. You need to collect the samples you will analyze.

1. Each student needs to obtain at least four test tubes with caps. A piece of wire mesh will already be inside the tube. Label each tube. These tubes will serve as nitrogen dioxide collection tubes.

2. Coat the wire mesh with 3 or 4 drops of triethanolamine (TEA) solution. TEA absorbs nitrogen dioxide. Rotate the tube to be sure the mesh is coated and then invert the tube over a waste beaker to allow the excess TEA to drain out. Immediately cap the tubes.

3. Store one of the tubes in your lab drawer or as directed by your laboratory instructor. This will serve as your control. Take the other tubes home with you. Remove the caps from the tubes and hang the tubes *upside down* at some site in or near your house. Attach them to a stake or tape them to a window or wall or fence post where they can hang undisturbed for a week. Note the exact time you hang the tubes.

4. Before you return to lab next week, recap the tubes and bring them back with you. Be sure you note the exact time you recap the tubes. Label each tube with the location at which it was hung (e.g., kitchen, back porch, etc.).

 Caution: The nitrosodiethanolamine produced when NO_2 reacts with TEA is a potential carcinogen. This compound should adhere securely to the mesh in the tubes. Keep the tubes in a secure place where they will not be mishandled by children and animals.

CALCULATIONS

Calculate the mass of phosphorus in the original mixture. Then express your result as a percentage of P_2O_5 in the original mixture. Calculate the percent by mass of the insoluble components in the original mixture. Obtain the expected % P_2O_5 and % insoluble components from your lab instructor. Calculate your % recovery.

$$\% \text{ recovery} = \frac{\text{experimental value}}{\text{expected value}} \times 100\%$$

REPORT

Your results should include sample calculations and a table of group and class results.

For your conclusion, summarize the procedure used to separate the water soluble components of the mixture from the water insoluble components of the mixture and the procedure used to separate the phosphorous from the water soluble components. Compare these two separation techniques. Explain whether or not each of the methods was effective. Support your conclusions with experimental data. Indicate which of the mixtures analyzed by members of the class contain an appropriate percentage of phosphorus for use as a fertilizer. Explain your choices.

Questions to Answer in Your Report

1. How do the percentage of P_2O_5 and the percentages of the insoluble components compare within your group? Are the results what you expected? Explain why or why not.

2. It is quite likely you did not obtain 100% recovery of the phosphorus or the insoluble materials from your original mixture. Some error is always expected whenever measurements are involved. Analytical chemists determine an acceptable range for the results of a given analysis dependent upon the accuracy of the equipment, glassware, and instruments involved in the analysis. If you are within 4% of the expected value, your error is within the acceptable limits. What factors could contribute to a % recovery greater or less than 100%? Be sure to consider the type of mixture you started with initially and the steps involved in each of your separations.

3. Was gravimetric analysis an appropriate technique to use in this experiment? Why or why not?

4. Which reactant was the limiting reactant in this experiment and which was in excess? What experimental evidence do you have to support your answer? Was it necessary to have one of the reactants in excess in this experiment? Why or why not?

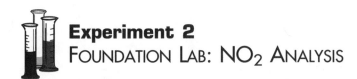

Experiment 2
FOUNDATION LAB: NO$_2$ ANALYSIS

Pre-Laboratory Assignment   **Due Before Lab Begins**

NAME: _____

Complete these exercises after reading the experiment but before coming to the laboratory to do it.

1. Based upon the chemistry involved in the analysis of NO$_2$ in the atmosphere in this experiment, explain why standard NaNO$_2$ solutions are used to prepare the calibration curve.

2. How many moles of nitrite ion are present if the cuvette contains 2.00 mL of 10.0 μmol/L NaNO$_2$ in a total volume of 3.00 mL?

3. Show that the units of C in the equation $Q = -\dfrac{DC\pi r^2 t}{z}$ are mol/cm^3.

4. The number of moles of absorbed nitrite that you measure in this experiment is distributed in the volume of air that passed under the sampler during the exposure period. Therefore the calculated concentration, C (see the equation in question 3), of NO$_2$ in the atmosphere is the *average* number of moles of NO$_2$ per cm^3 of air. If C is 2.63×10^{-12} mol NO$_2$ per cm^3 of air, express this concentration as a mole fraction (moles of NO$_2$/moles of air) and as parts per billion, ppb. Assume air is an ideal gas at 1 atm and 288 K.

5. Why is it recommended that you wear proper gloves during this experiment?

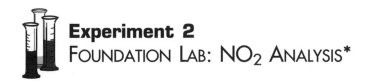

Experiment 2
FOUNDATION LAB: NO₂ ANALYSIS*

BACKGROUND

Today's atmosphere near the earth's surface is approximately 21% O_2, 78% N_2, and 1% Ar by volume. Trace gases, including CO_2, helium, neon, krypton, and xenon, are also present. Water vapor is present, but its composition varies on a day to day basis.

The nitrogen in the air is usually inert, but it can react to form nitrogen oxides (NO, NO_2, and N_2O—collectively known as NO_x's). This is not a favorable reaction under normal conditions. The simplest reaction between N_2 and O_2 forms NO, and this is highly endothermic. It only takes place to any appreciable event at high temperatures. The subsequent oxidation of NO to NO_2 is exothermic, however, and once formed, the oxides convert back to N_2 and O_2 very slowly. As a result, although NO_x's are only formed under special conditions, they are very persistent in the atmosphere.

There are important natural sources of NO_x's. Lightning and forest fires provide the high temperatures needed to initiate the reaction between N_2 and O_2 in the atmosphere. This essential process "fixes" the nitrogen in a form usable by plants. The NO_2 is washed from the atmosphere when it reacts with O_2 and H_2O to form HNO_3 molecules.

The high temperatures resulting from the combustion of fossil fuels also allow N_2 and O_2 in the atmosphere to react producing additional NO_x's. Transportation and stationary-source fuel combustion (power and industrial plants) are the primary anthropogenic (human-made) sources of NO_x's. Since many homes use gas appliances, the concentration of NO_2 indoors often exceeds outdoor concentrations. Finally tobacco burning may be a major source of NO_x indoors.

The nitrogen oxides, along with carbon monoxide, sulfur dioxide, particulates, toxic organic compounds, and volatile organic compounds, are air pollutants. NO_2 is an oxidant, it is soluble in biological tissue, and it is a respiratory irritant.[†] In addition, NO_x's, along with hydrocarbons, are involved in the production of photochemical smog. This smog, characterized by brown, hazy fumes, contains ground-level ozone, which causes respiratory distress and eye irritation in humans, degrades rubber, and destroys plants.

Environmental Monitoring

NO_x levels have stabilized over the past 20 years, due apparently to pollution control measures and increased energy conservation efforts. However, it is still important to monitor NO_x levels. Concentrations of pollutants, including NO_x, often vary throughout the day. Therefore, rather than analyze a single sample of air which provides an instantaneous picture of its composition, air is usually monitored over an extended period of time.

The simplest technique to monitor air over an extended period is to use a diffusion tube (see Figure J-2). This device—employed in this experiment—relies on the *passive* contact of its sampling agents with the air. To make sure the sampling is pas-

---

* This experiment follows the procedure outlined in Shooter, D. *J. Chem. Ed.,* **1993,** *70,* A133, A137–A140.

[†] When high concentrations of NO_2 are present in the atmosphere, an excessive quantity of HNO_3 may be produced, contributing to acid rain.

Plastic cap

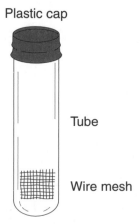

Tube

Wire mesh

FIGURE J-2 Diffusion Tube for Gas Sampling

sive, the tube must be mounted *upside down* so that the air that is sampled must slowly diffuse up the tube to reach the sampling agent. This also keeps wind and precipitation away from the sampling agent.

A simple equation relates the amount of air sampled per second by a diffusion tube with the tube's size and geometry:

$$Q = -\frac{DC\pi r^2 t}{z} \text{ mol} \qquad (1)$$

where Q = number of moles of gas absorbed (mol)
D = diffusion coefficient of the gas through air (cm^2 s^{-1})
C = concentration of the gas (mol cm^{-3})
r = internal radius of the tube (cm)
t = time the tube is exposed to air (s)
z = length of the tube (cm)

In this experiment, you will monitor NO_2 concentrations. You will know the time, t, of exposure, and you will determine the number of moles of NO_2 absorbed by your sample. For NO_2, the value for D is 0.154 cm^2 s^{-1}. You will then be able to calculate an average concentration, C, of NO_2. The negative sign indicates the flux is from high concentrations to low concentrations and can be ignored when reporting the concentration of NO_2.

The Chemistry of NO$_2$ Determination

The diffusion tubes you prepared at the end of the first week's experiment contained a piece of wire mesh which you coated with an aqueous solution of triethanolamine (TEA). TEA reacts very efficiently with NO_2 in the air to give a "nitrosamine" group:

$$\begin{array}{ccc} & CH_2CH_2OH & \\ N-CH_2CH_2OH & \xrightarrow{NO_2} & N-CH_2CH_2OH \quad + \text{ other products} \\ & CH_2CH_2OH & NO \end{array}$$

triethanolamine   nitrosodiethanolamine
TEA   NDELA

An assumption behind this experiment is that every molecule of NO_2 that comes in contact with the sampling agent (the TEA) is trapped on the wire mesh by this reaction. When you bring your tubes back to the lab, you will decompose the nitrosodiethanolamine by adding water in the presence of phosphoric acid. The key product is nitrous acid:

$$NDELA + H_2O \rightarrow R_2NH + HNO_2$$

The nitrous acid reacts in two steps to give a deep red complex according to the following equations. You will not be expected to know anything about these reactions, but for your information, they are a diazotization followed by a rapid electrophilic aromatic substitution.

You will determine the concentration of the dye spectrophotometrically. From this concentration, you will determine the number of moles of the dye and thus the number of moles of HNO_2, and finally the number of moles (Q) of the NO_2 that contacted the sampling agent.

> Caution: This lab requires the use of a sulfanilamide solution. It is potentially toxic and it is a potential mutagen. The use of proper gloves is advised. Wear appropriate safety goggles. Clean up all spills immediately with excess water. Be careful to avoid ingestion and, as always, wash hands carefully before leaving the lab. The nitrosodiethanolamine produced when NO_2 reacts with TEA is a potential carcinogen. This compound should adhere securely to the mesh in the tubes. Keep the tubes in a secure place where they will not be mishandled by children and animals.

PROCEDURE

Part I: Assign Tasks to Group Members

You will be working in a group of three people. Everyone is expected to participate in the experiment, to record data in his or her own notebook, and to complete his or her own report.

Part II: Preparation of Standard Solutions and a Calibration Curve

Your group needs to generate a calibration curve before you can determine the concentration of NO_2 in the atmosphere.

1. *Standard solutions.* Prepare a series of dilutions from the $NaNO_2$ stock solution. Use a 10-mL volumetric flask for each dilute solution.

2. *Cuvettes.* You have three reagents to add to each cuvette—one of the standard solutions and each of the two analysis solutions (sulfanilamide and NEDA). Your total volume cannot exceed the volume of the cuvette. Standard cuvettes hold approximately 3 mL. Check with your lab instructor for the volume of your cuvettes. Be sure you add the same amounts of each of the solutions to each cuvette. You want to make sure you have an excess of each of the analysis solutions. If you calculate the minimum quantity of sulfanilamide and NEDA you need to react with your most concentrated standard solution, then you know you will have an excess in all of your standards.

   After adding all reagents, *mix well*, then let the cuvettes stand for 15 min in order for the color to develop fully. Prepare a blank as well. A blank is a solution that contains all of the substances present in the samples being analyzed except the analyte (the substance of unknown concentration).

3. *Calibration curve.* Using a spectrophotometer, measure the absorbances of each of your solutions at 540 nm. See the Introduction for a review of the use of the Spectronic-20. Check with your laboratory instructor for instructions if you are using a different spectrophotometer. Graph absorbance vs. concentration and obtain the "best-fit" equation for the data (using a calculator capable of statistical analysis or a spreadsheet program.)

Part III: Analysis of NO_2^- Unknown

To check your calibration curve before analyzing your samples, each person in your group will obtain a nitrite solution of unknown concentration from your laboratory instructor. Treat the unknown exactly as you treated your standard solutions, and use your calibration curve to determine the nitrite concentration in the unknown. Check your results with your instructor. Compare your results with those of the other members of your group. As a group, decide if your calibration curve is accurate enough and, if it is not, repeat any aberrant measurements.

Part IV: Analysis of NO_2

You will be adding water and the two analysis solutions directly to the collection tubes, including the control tube. Add the same quantities of the analysis solutions as you used in the cuvettes when you prepared the calibration curve. Add enough water so that the total volume used with the samples is equivalent to the total volume used in the cuvettes with the standard solutions. Recap the tubes, shake them well, and let them stand for 15 min in order for the color to develop fully. Transfer the contents of the tubes to the cuvettes and measure the absorbances of the solutions.

Part V: Experimental Design for Week 3

You need to set up your soil analysis experiment this week to allow sufficient time for the separation of the mineral nitrogen from the soil. Set up your samples, the control, and the standard.

Instructions for this are given at the end of this week's procedure.

CALCULATIONS

Using your calibration curve and measured absorbances, determine the concentration of the nitrite ion in your unknown.

Using your calibration curve and measured absorbances, determine the concentration of the dye in your samples. Then calculate the moles of the dye, of HNO$_2$, and of NO$_2$. Using the number of moles of NO$_2$ and equation (1), calculate the concentration of NO$_2$ in the atmosphere. Express your answer in moles per liter. Assume air is an ideal gas with a total pressure of 1 atmosphere and a temperature of 290 K, and express your concentration as a mole fraction (moles of NO$_2$/moles of air). Finally, convert the mole fraction into units of parts per billion (ppb).

REPORT

Your report should include a summary of how you obtained the calibration curve and the curve itself (including the "best fit" equation and correlation constant). For the nitrite unknown, report the absorbance of the unknown nitrite solution and show a sample calculation of the concentration of the dye, the number of moles of nitrite, and the concentration of the nitrite in the unknown.

For the samples, prepare a table of the absorbance of the dye, concentration of the dye, number of moles of the dye, of HNO$_2$, and of NO$_2$, and the concentration of NO$_2$ in ppb.

Report the absorbance of the control and explain what information the control provided in this experiment.

Collect class data and draw conclusions regarding NO$_2$ concentrations based upon class results.

Sketch a map showing the locations where NO$_2$ samples were taken by the class. Is there a pattern? Are any of the samples above the U.S. Environment Protection Agency standard of 100 μg (micrograms) of NO$_2$ per cubic meter? Explain.

PREPARATION FOR THE APPLICATION LAB

In the application lab, you will use acid–base titration to determine the amount of mineral nitrogen in soil. Each student must set up four samples for analysis. Follow this procedure.

Most NH$_4^+$ ions in soils are attached to negatively charged sites on the edges of clay or organic matter particles. Potassium ions, K$^+$, bind to these sites more strongly than ammonium ions. Therefore, the ammonium ions can be released into solution by treating the soil sample with a KCl solution. Nitrate ions present in the soil are reduced to ammonium ions by the addition of a reducing agent known as Devarda's alloy. This alloy is a mixture of metals consisting of 50 parts Cu, 45 parts Al, and 5 parts Zn. Then, by making the solution basic, the NH$_4^+$ ions are volatilized from the solution as ammonia, NH$_3$.

The ammonia volatilizes and diffuses slowly into the air. A boric acid (H$_3$BO$_3$) solution, kept in proximity to but not in direct contact with the soil sample, will absorb the ammonia. An indicator added to the boric acid solution changes color as the NH$_3$ is absorbed. Prepare four small beakers or specimen cups with about 15 mL of the boric acid–indicator solution. Also, obtain four screw-top jars. Make sure they are clean and dry.

In the first jar, place 2 to 3 g of fresh undisturbed soil. Record the mass of the soil used to ±0.001 g. In the second jar, place 2 to 3 g of fresh disturbed soil. Record the mass of the soil used to ±0.001 g. To *each* of your soil samples, add about 20 mL of 2 M KCl solution, sprinkle about 0.2 g of Devarda's alloy, and add approximately 0.2 g of solid MgO (to raise the pH). Neither the Devarda's alloy, nor the MgO need to be weighed accurately. Swirl to mix.

Place one of the boric acid–indicator samples in each of the jars, then seal them.

A control must also be setup in this experiment. The control jar is identical to the sample jars *with the exception of the soil*.

It is important to check your experimental design for complete recovery of mineral nitrogen by analyzing a standard solution (a solution in which the quantity of mineral nitrogen is known). Set up another jar in which you replace the soil with 2.00 mL (measured exactly) of the standard solution. Swirl to mix.

The systems must sit undisturbed for a week. Place all of your jars in a location specified by your instructor.

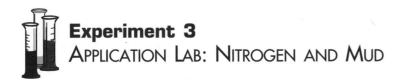

Experiment 3
APPLICATION LAB: NITROGEN AND MUD

Pre-Laboratory Assignment   **Due Before Lab Begins**

NAME: _____

Complete these exercises after reading the experiment but before coming to the laboratory to do it.

1. You will titrate a boric acid solution that has absorbed the ammonia from a soil sample. It is not necessary to know the exact volume or concentration of the boric acid. Explain why not. (*Hint:* Combine equations 1 and 2 in the laboratory procedure).

2. To prepare for this week's lab, you added a KCl solution, Devarda's alloy, and MgO to your soil sample in order to analyze for mineral nitrogen. Sketch a jar containing soil and indicate how the nitrogen is initially present in the soil. Then draw three more jars showing the impact of each of the three substances, KCl, Devarda's alloy, and MgO, on the soil sample.

3. a. Calculate the number of moles of ammonia absorbed by a boric acid solution if 3.25 mL of 0.0025 M H_2SO_4 were needed to titrate the solution to the end point. (Assume both hydrogens of the sulfuric acid react.)

    b. From the number of moles of ammonia calculated in part (a), determine the number of micrograms of nitrogen per gram of dry soil. Assume you have 2.753 g of soil and all of the ammonia diffused from the soil.

4. If some of the acid solution drips down the side of the stock bottle as you fill your own beaker, what should you do?

Experiment 3
APPLICATION LAB: NITROGEN AND MUD*

BACKGROUND

Nitrogen and phosphorus are the two nutrients that most often limit the growth of plants. For this reason much effort has been put into finding ways to measure nitrogen and phosphorus in soil and water. Nitrogen is present in soil and water in both organic compounds (primarily in amino acids) and inorganic compounds (as ammonium, NH_4^+, and nitrate, NO_3^-, ions). The nitrogen present in inorganic compounds is known as mineral nitrogen. Microorganisms, in a process known as nitrification, rapidly catalyze the oxidation of ammonium ions first to nitrite ions (NO_2^-) and then to nitrate ions (with some N_2O as a byproduct). These microorganisms compete for the ammonium ions with roots in soil and with algae in water.

You will be determining the percent by mass of mineral nitrogen in two samples of soil, one that has been disturbed and one that is undisturbed, through an indirect titrimetric method. Thus you will measure the amount of nitrogen available for a plant root, bacterium, or fungi to take up and you can determine the impact of the disturbance on the mineral nitrogen content. To analyze for mineral nitrogen, you need to separate the mineral nitrogen from the rest of the sample. The method you will use to accomplish this separation involves the following steps:

1. Displacement of NH_4^+ and NO_3^- ions from the soil by KCl
2. Conversion of NO_3^- to NH_4^+ with alloy
3. Volatilization of the NH_4^+ as NH_3
4. Absorption of the resulting NH_3 vapor by a boric acid solution
5. Titration of the resulting solution with an acid

NOTE: This experiment must be set up one week prior to the analysis to allow time for the separation of the mineral nitrogen. You also need to dry a separate portion of your original fresh soil sample in a drying oven so that you can report your results per gram of dry soil. Weigh a 2- to 3-g sample of wet soil, record the mass to 0.001 g, and place it in the drying oven.

 Caution: You will be titrating with a dilute solution of a strong acid during this experiment. Clean up small spills (several drops) with excess water. Neutralize large spills (several milliliters) with sodium hydrogen carbonate or as indicated by your instructor.

PROCEDURE

In the samples you set up last week, the mineral nitrogen, present as ammonium and nitrate ions, has been volatilized into the atmosphere of the jar as ammonia gas. The boric acid solution in the jar absorbs this ammonia. An indicator added to the boric acid solution changes color as the NH_3 is absorbed.

---

* The method for determining the mineral nitrogen content in soil utilized in this experiment is described in Mulvaney, R. L., "Nitrogen—Inorganic Forms," pp. 1123–1184 in D. L. Sparks et al., editors, *Methods of Soil Analysis*. Part 3, Soil Science Society of American Book Series 5. ASA and SSSA: Madison, WI, 1996.

$$H_3BO_3 \ (aq) + NH_3 \ (g) \rightarrow H_2BO_3^- \ (aq) + NH_4^+ \ (aq) \qquad \text{(1)}$$

The indicator is a mixture of bromocresol green and methyl red. Record the initial color of the mixed indicator in the boric acid solution. The indicator will turn green as the solution absorbs ammonia.

The resulting $H_2BO_3^-$ can then be titrated with an acid solution back to the original color of the indicator by an acid that provides hydrogen ions.

$$H^+ \ (aq) + H_2BO_3^- \ (aq) \rightarrow H_3BO_3 \ (aq) \qquad \text{(2)}$$

The amount of acid needed to reach the end point is used to determine the amount of nitrogen present in the original sample. This is known as an indirect method because the ammonia, the compound of interest, is not directly titrated.

The analysis for nitrogen is completed by titrating the boric acid solution that absorbed the ammonia from your samples, the control, and the standard.

A pair of students should practice this titration together, but each student will be responsible for analysis of the four samples prepared last week.

To practice the titration, place about 20 mL of the boric acid solution into a beaker. Use a volumetric pipet to add a known amount of standardized ammonia solution to the beaker, and swirl gently to mix. Note the color change. Titrate this solution with standardized acid to the end point. This should be done at least twice to be sure that the end point is reproducible. Then you are ready to titrate each of the four experiment set-ups.

> **NOTE:** You dried a separate sample of the same soil you are analyzing for mineral nitrogen. You needed to do so in order to determine the moisture content of the soil. Your results of the analysis for mineral nitrogen are to be reported in micrograms of nitrogen per gram of dry soil. The sample you will analyze for mineral nitrogen is not dry, but if you know the percentage, by mass, of water in an equivalent soil sample, you can calculate the dry mass of the sample used in the analysis.

CALCULATIONS

Calculate the percentage (by mass) of water in your soil sample.

Calculate the dry mass of the soil sample used in your analysis for mineral nitrogen.

Calculate the number of moles of acid used in your titration involving the sample and the control. Use this value to calculate the number of moles of ammonia absorbed by the boric acid. Finally, calculate the number of micrograms of mineral nitrogen present in your soil sample. Be sure to factor in the control. Report your results as micrograms of nitrogen per gram of dry soil.

Calculate the number of moles of acid used in your titration involving the standard. Use this value to calculate the number of moles of ammonia absorbed by the boric acid. Finally, calculate the experimental number of micrograms of nitrogen per milliliter of solution in the standard. Obtain the expected value from your laboratory instructor and calculate your percent recovery.

REPORT

Tabulate all results and show sample calculations.

Describe the experimental design used in this experiment and comment on the validity of the design to accomplish the goal of the experiment. Explain the role of the control in this experiment. Explain how you were able to calculate the mass of dry soil used in the analyses.

Share the results of your soil analysis and your percent recovery of the nitrogen in the standard with the entire class. Draw conclusions regarding the effectiveness of the method used in this experiment for analysis of mineral nitrogen. Also discuss the impact of the disturbance on the mineral nitrogen concentration in the soil. Support your conclusions with data.

If your percent recovery of mineral nitrogen in the standard solution is not 100% ± 2%, explain what factors may have contributed to the recovery error.

QUESTIONS TO ANSWER IN YOUR REPORT

What about the original question you couldn't answer in the Scenario? If you build a cabin in the forest, will the disturbance of construction and regular use raise or lower the ammonium and nitrate concentrations in water and soil? Answer the scenario question based upon your responses to the following questions. Include your answers to all questions in your final report.

1. Does disturbance of soil increase or decrease the concentration of ammonium or nitrate? Use your experimental results to support your answer.

2. If a soil is disturbed for years, what is likely to be the effect on soil fertility? Explain.

3. Atmospheric deposition of important ions is recorded by the National Atmospheric Deposition Program (NADP). Look for their Web site at *http://nadp.sws.uiuc.edu*. Using isopleth maps of ammonium and nitrate at the NADP Web site, look at maps of ammonium and nitrate deposition. Where are the centers of ammonium deposition? Where are the centers of nitrate deposition? Can you infer sources of ammonium and nitrate in our atmosphere from the maps? Explain.

INDEX